A TREATISE

ON THE

EMPLOYMENT OF LIGHT TROOPS

ON

ACTUAL SERVICE;

CONTAINING

GENERAL PRINCIPLES, COMPILED FROM EMINENT
PRACTICAL AUTHORS,

British and Foreign;

ILLUSTRATED BY NUMEROUS EXAMPLES,

SHOWING THEIR APPLICATION TO SKIRMISHING IN THE FIELD,

IN CONFORMITY WITH

Her Majesty's Regulations.

By Lt.-Colonel CHAS. LESLIE, K.H.,
Late 60th King's Royal Rifles.

The Naval & Military Press Ltd

published in association with

ROYAL ARMOURIES

Published by the
The Naval & Military Press
in association with the Royal Armouries

Unit 10 Ridgewood Industrial Park,
Uckfield, East Sussex, TN22 5QE
Tel: +44 (0) 1825 749494
Fax: +44 (0) 1825 765701

For a full listing of all N&MP titles, visit:
www.naval-military-press.com

MILITARY HISTORY AT YOUR FINGERTIPS

Online genealogy research:
www.military-genealogy.com

ROYAL ARMOURIES

The Library & Archives Department at the Royal Armouries Museum, Leeds, specialises in the history and development of armour and weapons from earliest times to the present day. Material relating to the development of artillery and modern fortifications is held at the Royal Armouries Museum, Fort Nelson.

For further information contact:
Royal Armouries Museum, Library, Armouries Drive,
Leeds, West Yorkshire LS10 1LT
Royal Armouries, Library, Fort Nelson, Down End Road, Fareham PO17 6AN

Or visit the Museum's website at
www.armouries.org.uk

In reprinting in facsimile from the original, any imperfections are inevitably reproduced and the quality may fall short of modern type and cartographic standards.

ERRATA.

Page 11, No. 2, in last line but one, *for* press, *read* pass.
 „ 50, No. 26, last line, *for* No. 11, *read* No. 4.
 „ 69, Top heading, *for* Rear, *read* Front.
 „ 131, No. 58, first line, *for* Lead, *read* Head.
 „ 144, No. 113, line 11, *for* on, *read* in.

PREFACE.

IN offering this Treatise to the notice of my comrades in arms, I may be permitted to premise that I have no *new System* to propose—no *new Theory* to establish—no new *Drill* to introduce—nor, in fact, anything that would impose on the soldier a burden to learn.

The arduous and protracted struggle in which Europe had to contest against the great military genius of the age (and which, notwithstanding that his celebrated campaigns led in the first instance to the subjugation and overthrow of empires and kingdoms, eventually terminated in his discomfiture and downfal) has tended to promote great improvements in every department of military science, tactics, &c., and in none more so than in the system of organization and employment of Light Troops.

In the various vicissitudes of defeat and victory which the Continental armies experienced during that eventful period, the great importance and utility of Light Troops became so conspicuous that the subject has since then deeply engaged the attention of all military powers—and none more so than our former gallant opponents the French, who now admit that great defects existed in their former system of organization, and in the practical employment of their Light Troops in the field (*See* Note 1). That whatever success their Light Troops acquired was solely accomplished by force of numbers, swarms of Skirmishers being thrown out in overwhelming force, consequently at an immense sacrifice of their bravest men, while many of their adversaries, more experienced in this branch of the service, frequently attained, by their skill, tact, and dexterity, many important advantages with far inferior numbers (*See* Note 2), and that even in their career of victory they always encountered the most determined opposition in those countries where the inhabitants were accustomed to the use of the rifle, or where the " Levées en masse" acted as Light Troops—such as the Tyrolese, the Swiss, Spanish Guerillas, Cossacks, &c.

With regard to our own gallant Light Corps, it may be observed that, although the high state of efficiency is still maintained which these corps acquired during the splendid campaigns under the illustrious Wellington in the Peninsula, where the glorious example of stern resistance shown by the British arms in their victorious career first roused other nations to shake off the galling yoke of their oppressor, and which was finally accomplished by British prowess and Prussian valour at Waterloo, it must be recollected that the numerous skilful and experienced officers then formed are now fast disappearing from the service, and will shortly be extinct; and that, in consequence of the lengthened peace in Europe which the nation has enjoyed, the officers who have since entered the service cannot have the opportunity or advantage of practically learning this branch of their profession in the field as their elder brethren in arms had. Hence without great application and study only a theoretical knowledge of what is truly required on actual service can be obtained. During the course of the vast operations of the Continental and our Armies, much additional practical experience and information in this particular branch of the service had been acquired, many rules and practices which were no longer applicable to the more recent modes of carrying on this sort of warfare were abrogated, and many defects which existed in former systems have been rectified; in fine, almost every power, whether in the old or new world, have of late not only greatly increased this arm, but have all more or less revised their former systems so as to place their Light Troops on a more efficient footing, not more in regard to training and discipline than that they should excel in every particular; and to further this end all the late improvements in fire-arms, equipments, &c., have in most cases been adopted and introduced.

This subject has drawn forth the abilities of many authors of great merit, in Germany and Prussia in particular. France has likewise produced several valuable writers; indeed, many distinguished Continental officers have, since the period of the war, devoted their talents in recording and illustrating the experience then acquired, all demonstrating the increased advantage which may be derived from the proper employment of Light Troops, while we have few or no native writers who enter at large on this interesting topic.

Having served long in a rifle corps, and taking a great interest in this branch of the profession, my attention has been directed to acquire every possible information connected with the subject. In making a military tour to view the Continental armies, and to prosecute my researches on this head, I found that most of the countries I visited possessed various authors of great repute treating on the system of training Light Troops. The employment of these in the field, and in partizan warfare, affording an abundant source of authentic practical information, I consequently procured at Berlin, Vienna, Hanover, Paris, &c., those works which were esteemed the best, with, also, the Army Regulations of the different countries; from these have been selected all such general principles and practical details which the various writers seemed most agreed upon as having been proved by experience to be the most eligible for all purposes in the field. Having likewise obtained much valuable information from the work of the gallant Lieutenant-Colonel Leach, late of the Rifle Brigade, " Recollections and Reflections relating to the Duties of Light Troops composing Advanced Corps," and also from our eloquent military historian Napier. A compilation of *practical facts, the fruits of experience,* has thus been acquired, drawn from military authorities who had served and studied during the grandest scale of memorable operations in this or any other age.

My former work on the Application of Light Drill to skirmishing in the Field being all disposed of, and it having been intimated to me that a second edition would be desirable, I determined to revise that work, and to introduce all the recent information I had obtained, and to bring the whole into strict conformity with Her Majesty's Regulations. With this view, and bearing in mind the paragraph, Part V., page 312, relating to field training to be practised by battalions of Light Infantry, viz.—" It is not contemplated, nor would it be practicable
" to lay down specific rules for the guidance of command-
" ing officers in the practical course of field-discipline here
" prescribed. Upon their *judgment* and *experience* the
" proficiency of their corps in individual expertness must,
" from the nature of the instruction, mainly depend. All
" therefore that can be desired on this point is, that the
" subject should be *classed* under *certain heads,* with a
" few accompanying remarks pointing out the chief objects
" to be held in view, so that an *uniform system* may pre-
" vail, varying perhaps a little in the method of teaching,

"but still arriving, although by different roads, at the "same end." It appeared to me that, in accordance with the latitude thus permitted to commanding officers, the intentions above expressed could not better be fulfilled, and at the same time promote in some degree the desired end, that an uniform system might prevail in the practical course of field operations, than by collecting the general practice of our own most distinguished Light Corps, whose proficiency in field duties was acquired and established by the *judgment* and *experience* of *active* and *intelligent commanding officers* during their gallant operations, in a series of brilliant campaigns in the field. Hence such practice has been arranged in combination with the General Principles, &c., &c., and brought into harmony with Her Majesty's Regulations, from which all the sections of Part V., relating to Skirmishing, have likewise been embodied at such parts where they had reference, or to which the principles there laid down applied, and the whole *classed under certain heads*, so that although the arrangement may be different, the whole system of the *Established Light Drill* will be found entire, this work being solely intended to illustrate more fully the principles and practice therein contained in their application to the various operations of attack and defence when acting in the field.

The general principles and maxims, with the examples quoted, have reference in almost every instance to some event which actually occurred on many momentous occasions in the battles or affairs of out-post which took place between the various contending armies during the great theatre of war. Many of them serve to illustrate the operations, and to record the achievements of our gallant army and distinguished Light Corps.

I now commit this work to the indulgence of my

military brethren, in the hope that it may obtain the approbation of the elder, and that it may prove of utility to the younger branches serving in Light Corps, to whom I trust it will afford a general *outline* of the various arduous duties they will be called upon to perform, and of the qualifications they will be expected to display, in the event of their being employed on actual service in the affairs of out-post.

In order to render a work of this kind more complete, there ought to have been (had time permitted) an additional Part, treating on Advanced Post Duties, Pickets, Patroles, &c., as conducted during the Peninsular Campaigns,—Reconnoitring,—Advanced Corps, whether to cover Winter Cantonments, an Army of Occupation, or to observe the Enemy (the notice of such corps in Section XVIII. of this work being merely a sketch of their general object and duty).

The various enterprises and operations of Partizan and Guerilla warfare, such as Escort Duty, Protecting Convoys, Foraging Parties, &c., Ambuscades, Surprising Posts, Cutting off Detachments, Convoys, &c., Detached Parties acting on an Enemy's Flank or Rear, &c.

Hence young Officers would do well to study the best works on such subjects, and all narratives of war relating to them.

THE AUTHOR.

Brighton, October, 1842.

AUTHORITIES CONSULTED.

NUMEROUS authorities have been consulted, both Foreign and British. Amongst the former the following may be mentioned as those from whom much practical information has been obtained:—

Der Kleine Krieg (Partizan Warfare), by Colonel Decker, Prussian Service; now translated into French, and held as a standard work.

Captain Petre, Prussian Service, on Jägers, and the Movements of Light Troops.

Captain Koester's Hand-Buch.

The Prussian and Hanoverian Army Regulations, and various other works.

Instruction Pratique sur la Guerre de Campagne, à l'Usage des Officiers de l'Armée Autrechenne. Vienna, 1816.

Campagne de 1799 en Allemagne et en Suisse. Vienna, 1820.

Traité Théorique et Pratique des Opérations *Secondaires* de la Guerre. Par M. A. Lallemand, Chef de Bataillon du Corps d'Etat Major.

Monsieur le Baron de Chambrun, Colonel of 4th Regiment French Light Infantry, on the Movements of Light Troops in Skirmishing Order.

Colonel Marnier, French Service, on the Amelioration of the system of Light Troops.

Captain de Beurman, of the 55th French Regiment of the Line, on the better Organization of Light Troops.

Lieut.-Colonel de Beauval on Tirailleurs.

Captain Gustave Delvigne, French Service, on Rifles and Corps of Sharp-shooters in Africa.

Captain de Forestier, on Light Infantry.

The French Regulations for Light Infantry.

The French Instructions for Light Troops in the Field.

The French Manual for Light Cavalry and the Chasseurs à Cheval.

Monsieur Pinette, on the Bayonet Exercise to resist Cavalry.

With regard to British authors, it will be observed that I have drawn largely from Lieut.-Colonel Leach's excellent little work, likewise from Major General Napier's History. I am also indebted to several correspondents in the United Service Journal for much useful information on various subjects. I feel assured that all these authors will feel more flattered than displeased at the liberty I have taken in availing myself of their suggestions. I trust that they will accept of the general acknowledgment: particular quotations have in general been omitted; it would have swelled this volume to too great a size to have repeatedly particularized each author, or to have given chapter and verse for every event in numerous various places. Indeed my aim has been to compress the whole into a portable manual of convenient size, and to reduce the price to the lowest possible scale. For these reasons, explanatory plates have not been given at present. It is, however, in contemplation to publish a set at a very moderate rate.

NOTES.

EXTRACTS FROM FRENCH AUTHORS.

Note 1.—Captain de Beurman, of the 55th French Regiment of the Line, in his work on the Better Organization of Light Troops, says:—
" I cannot comprehend how *we French*, who occupy so high a place with
" regard to military instructions and glory, should remain in rear of all
" our neighbours in the formation and organization of our Light Troops.
" Feeling the necessity there is for establishing them on a better footing to
" enable them to display all the advantages which they offer to the Army,
" on whom they exercise so great an influence. I have not hesitated, in
" conformity with the wishes of almost all officers who have studied the
" art of war, to state my views, and which are in accordance with those
" of most military authors who have treated on this arm. There ought
" to be no delay in giving a *new* organization to our regiments of Light
" Infantry, who at present only bear the name, and who are far removed
" from the true principles of their primitive institution. The security of
" an army, the correctness of its measures, which tend so decidedly to pro-
" mote its success, depend immediately on the superior vigilance, in-
" struction, and qualifications of the Light Troops compared to those of the
" Enemy. It is therefore urgently necessary that we should re-organize
" our Light Infantry, and also the instructions for their duties in the
" field, so that they may be able to prove themselves superior to those of
" the Enemy. All our neighbours have their Light Infantry all organized,
" more or less, on the proper principles of their institution. It is only
" we, the French, who, in spite of the lessons of the past and the necessity
" of the present, have not yet understood all the importance of a special
" and more complete organization for this arm, page 18. In war we
" find the necessity to cover the front of our line with Skirmishers, to
" clear the march of our columns by Troops who can search the country,
" so that, in default of others, we were obliged to dedicate the Light Com-
" panies of regiments and *even Battalion* Companies to this service. But
" these soldiers can only imperfectly execute their duty, because they
" have only learned this branch of their profession but very imperfectly.
" It consequently happens that these very soldiers, when employed as
" Skirmishers, often receive contradictory orders to what they have been
" accustomed to, which confuses their ideas. In this manner we spoil
" the soldiers of the line, in requiring them to quit their ranks, while the
" habit of manœuvring in close order renders them heavy and awkward
" as Light Troops."

" The wars of the Revolution, in giving to the military art so brilliant a
" flight, revealed the great utility of Corps of Tirailleurs. The different
" armies of 1793, which rose as if by enchantment, formed in haste from
" a ' Levée en masse,' had only boiling courage to oppose to the *address and*
" *dexterity* of the *Austrian Sharp-shooters*. Our Generals felt the necessity
" of promptly remedying our then military inexperience by the formation

"of Free Corps, and exercising our Light Infantry as Tirailleurs. Hence
"in those times this sort of fighting became much in favour, indeed
"many important battles were crowned with brilliant success by the
"attack 'en Tirailleur.' Besides the Light Infantry regiments, the
"French Free Mountain Legion, those of Rosenthal, of Westerman, and
"others, rendered important services, and showed of how great impor-
"tance it would have been to have *reconstructed permanently our Light
"Troops* on a *good system*. But no, all these Corps, who only appeared
"for an instant on the theatre of war, were either dismissed or were
"amalgamated with other corps when the *regular armies* became more
"perfectly organized. We never thought of imitating the excellent
"example of our neighbours, who had well organized Light Corps, and
"which afforded them often great success against us. Indeed, after-
"wards, although various Battalions of *Chasseurs*, such as those of the
"Alps, of Corsica, of the Pyrenees, &c. &c., were always at the head of
"our columns, they did not differ from the Infantry of the line except
"in their uniform, for their arms, equipments, and drill were in every
"respect exactly the same. When Marshal Gouvion St. Cyr became
"Minister of War, finding the necessity of dividing the Infantry into two
"distinct kinds, he ordered Legions of Chasseurs to be formed: he
"changed the uniform, equipments, and arms, but *neglected to regulate
"the Military Instructions requisite* for this *branch* of the service. Since
"then these Legions have been transformed into Regiments of the Line
"or into self-styled Light Infantry ones, *without attaining* the *object* for
"*which they are destined*. If these corps had been kept up, however de-
"fective their organization might have been, ameliorations might have
"been introduced to render them effective Light Troops. And *we* should
"*not* have *experienced such rude Schools* of *Adversity* in the *Wars* in *Tyrol*
"and *Spain*, where the want of good Light Troops was most severely
"felt.

"It is true we have in the French army companies of Carabiniers, but
"these are only nominally so, they being composed of ordinary Grena-
"diers, good and chosen men, it is true, but men not accustomed to the
"use of the arm of which they bear the name."—*See* Colonel Marnier's
work on "The Better Organization of Light Troops," page 10.

Le Baron de Chambrun, Colonel of the 4th French Regiment of
Light Infantry, in his Treatise on the Manœuvres of Troops in Skirmishing order, states, that "He had only written his little work and
"thrown out some hints in the expectation that some abler hand would
"produce a complete work for the Manœuvres of Light Troops. Instruc-
"tion on this head is much wanted in the Infantry regiments, all the
"present (French) works on Light Troops are very unsatisfactory and
"very defective."

NOTE 2.—Captain Gustave Delvigne, in his work on Rifles and Corps
of Sharp-shooters in Africa, states:—"That previous to going on the
"expedition to Algiers, where they expected to engage in a sharp-shooting
"warfare against men reckoned most skilful in the use of their arms, he
"hesitated whether he might not have been deceived in having too strongly
"expressed his opinion against the defective system of instruction for the
"French Voltigeurs. But that, from what he had there seen, it justified
"him in his conviction that an amelioration in this branch of the service
"was indispensibly necessary; that every military person of common
"observation will agree that the object proposed in the formation of the

"Light Troops had failed in a great degree, owing to the want of a pro-
"per system, whether in regard to their instruction or armament. With
"regard to our Companies of Voltigeurs, it is certainly very fine to see so
"many smart fellows formed on parade; but with regard to the many other
"essential qualities required for men destined for this light service, they
"are seldom considered in selecting the men—all is sacrificed to appear-
"ance and good looks. But when war occurs, when these companies are
"called upon to fulfil the conditions for which Voltigeurs are instituted—
"when it is necessary to display tact and intelligence, and to use their arms
"—not to make a show by noise and smoke, but to hit the Enemy—
"then, but a little too late, our eyes are opened, and we are forced to sub-
"mit to the consequences of so vicious an organization. Then in place
"of being able to cover the approaches of a defensive position with good
"Voltigeurs or Riflemen, we have no other alternative left in order to
"keep off the experienced Light Infantry of the Enemy than to throw out
"*whole Battalions* as *Skirmishers*, throwing away thousands of balls, of
"which only a few perhaps strike an Enemy, while our loss of men is
"most enormous. We then find out that the Enemy's Skirmishers get
"accustomed to the whizzing of such a prodigious quantity of balls,
"which do them but little injury. They get emboldened — they
"approach close up—skilfully taking advantage of cover, and occasion so
"much more mischief, because in place of a small number of well-
"instructed men firing with coolness from behind good shelter, it has
"been necessary for us to oppose to them *tenfold numbers*, who, *either*
"from *ignorance* of *selecting* cover, or the difficulty of so many men find-
"ing shelter, are obliged to be *exposed* to the shots of the Enemy.

"The war in Africa, which at the commencement was, with few excep-
"tions, a guerilla warfare, demonstrated this truth in the most particular
"manner to all those who are animated with a desire that we should
"profit by experience to ameliorate this state of things; all will agree
"that our Light Infantry are far from knowing how to draw from their
"arms the best possible effects; to demonstrate this one fact is sufficient,
"viz. :—That in the space of only 15 days three millions of cartridges
"were expended almost entirely by the Tirailleurs alone. The mounted
"Arabs galloped up in succession to our line of Skirmishers, fired, and
"retired, while those on foot stole on with great address through the
"bushes, &c., &c., until within a good range, and proved to our loss their
"great practical dexterity in the use of their arms, while our Tirailleurs
"could only answer them by a prodigious number of balls, but of which
"the greater part scarcely reached them, and did little execution,
"because, from the distance the Enemy kept, and their being well under
"cover, a ball hitting any of them was a matter of chance. Those who
"seek to ascertain the true cause may ask, what, on one part, gave us
"such a decided superiority over the Arabs in regular pitched battles,
"so that in spite of superior numbers we always routed them with little
"or no loss, and on the other hand gave us so little advantage, if not
"inferiority, in Skirmishing warfare. Did we not know the contrary, it
"might have been attributed to the excellence of their fire-arms and
"superiority of their range; such however was not the case, their arms
"were much inferior to ours, being very imperfect, and of various sorts
"and sizes, without any uniformity. Hence it can only be attributed to
"their greater skill as marksmen, and their dexterity in making the best
"possible use of bad arms, and the advantages of cover, that the Arabs

" owed their superiority over us. It occasions chagrin to think that
" our Light Infantry, so well disciplined, and, in other respects so
" well instructed and better armed than the Arabs, were nevertheless
" inferior in Skirmishing, arising, unquestionably, from the want of a
" *proper system* of *practical instruction.*"

Captain Delvigne afterwards adds :—" The recent formation of Batta-
" lions of Light Troops prove that military persons of great experience
" and abilities acknowledge the necessity of a better organized system of
" Sharpshooters, and that this conviction of the utility of a superior
" description of Light Infantry had alone determined them to *rise above*
" the *prejudices* of a *false routine* so long hitherto persisted in."

It may be mentioned that since the authors above quoted wrote on the
subject, the French government have taken up the question relating to
the introduction of a better system for the organization, &c., &c., of their
Light Troops with great energy, urged no doubt to take this step in consequence of the perseverance with which these gentlemen persisted in
pointing out the errors of the former defective system, the zeal and
ability they displayed in demonstrating a more improved one, and the
necessity of its adoption. No less than ten new Corps of Light Infantry
have been raised in 1841. They were formed and trained at St. Omer,
under the personal direction of the late gallant Duke of Orleans, who
devoted much attention to the subject, and caused all the most approved
methods to be adopted, such as were proved by practice and experience
in the field to be requisite. Indeed the soldiers of the New Rifle Corps,
the Chasseurs d'Afrique, styled Chasseurs d'Orleans, are said now to
equal any in Europe.

In a series of experiments lately made at Vincennes, in the presence of
the Duke of Montpensier and General Rostolan, as to the relative
merits of the muskets used by the Infantry of the Line, and the improved
Carabin 'Rayée (rifles), invented by M. Delvigne, some of the best
marksmen of the 68th Regiment of the Line, only put 7 balls out of
200 into the target at 400 yards distance, whilst the Chasseurs placed
10 times the number, in the same number of shots, at the same distance.
They likewise put in 33 balls out of 200 at 500 yards distance, and 25
at 600 yards. As these ranges exceed the usual practice of ordinary
rifles, it tends to prove the superiority of the improved arm now adopted
in the French Light Corps.

CONTENTS.

	Page
INTRODUCTION	xxi

SECTION I.

Utility and Employment of Light Troops	1
General Purposes	2
Application of Skirmishing	2

SECTION II.

System of Training, and Qualifications required	3
Mechanical and Intellectual	3
Choice of Men	3
Qualifications, Marksmen, &c.	4
Taking Advantage of Ground	5

SECTION III.

Mode of Instruction	6
Athletic Exercises	8
Riflemen	8
Regiments of the Line	9

SECTION IV.

Duty of Officers and Non-Commissioned Officers	11
Commanding Officers	12
Captains	13
Subalterns	13
Sergeants	14

SECTION V.

Ground, Varieties of	16
Positions—Defensive—Offensive	17
Retreating	17
Occupying Positions	18
Cover, Varieties of	18
Occupying Cover	19

CONTENTS.

Page

SECTION VI.

Movements of a Line of Skirmishers	20
Time of Movement	20
Changes of Position	21
Point of Direction	21
Dressing	21
Unity of the Line	21
Not to collect in Groups	22
When the "Halt" is sounded	22
General Rule	23

SECTION VII.

Skirmishing Order	24
Disposition to Engage—General Rule :—	
Skirmishers and Supports	25
Skirmishers, Supports, and Reserve	25
Distances of these Bodies	26
Unity of Companies	26

SECTION VIII.

Mode of throwing Troops into Skirmishing Order	28
Strength employed	28
Extending—On the March, or when Halted	29
Distance of Skirmishers and Supports	29
Between Files	29
Point of Direction	30
Companies thrown out from a Battalion	30
From Line	30
From Column	31
To mask a Retreat	33
Skirmishers conform to Movements of Battalion	34
When Battalion form Squares	34
Clearing the Front	34
Recalled—By Close—Assembly	35
Remarks	35
A whole Battalion to Skirmish	36
Supports and Skirmishers	36
From Line	36
From Column, to Deploy and Extend	37
Battalion Reformed	38
A Body thrown into Light Order	38
Skirmishers, Supports, and a Reserve :—	
From Column	39
From Line	39
Supports and Reserve conform to Skirmishers	40
Battalion Reformed	40

CONTENTS. xvii

Page
SECTION IX.

Skirmishers Firing in Action 42
    General Observations 43
        Range, Position in Firing, &c. 44
           Simultaneous Attacks 45
                Example 45
    Firing Halted, in Cover 46
    Firing Advancing, in General Line 47
        Working way through Cover 47
        Loading Halted, or on March 48
    French Mode 48
    Firing Retreating—General Line 49
        Alternate Lines 50
        Alternate Ranks 50
    French Mode 51
    Firing taking Ground to a Flank 51

SECTION X.

Duty of Supports, Utility of 53
    Strength 53
    Position 54
    Number of, Distance from Skirmishers . . . 54
        Conform to Skirmishers 56
    Preparing for Cavalry—Squares 56
    Duty during a Retreat 57
    New Lines to a Flank 58
    Reinforcing Skirmishers 58
    Attacking in Close Order 59

SECTION XI.

Duty of Reserve 60
    Position in Rear 60
    Conform to Skirmishers 61
    Composition of 62
    Local Reserves 62
        Flank Parties 62

SECTION XII.

Internal Changes and Combined Movements . . . 63
    Relieving Skirmishers—Halted 63
        Advancing—Retreating 64
    Reinforcing Skirmishers 65
        Line Strengthened—Prolonged . . . 65
    Chain Weakened—Diminished 65
    Changing Front 66
        By Wheeling 66
        By Shoulders Forward 66
    Closing, and Forming Line to a Flank . . . 67
    New Line to a Flank 68
        Direct—Oblique 68

xviii CONTENTS.

Page

SECTION XIII.

Rallying—Recovering Order 70
    Skirmishers Recalled 71
    Resisting Cavalry 72
        Squares 73
            Examples 75
        Different Modes of Forming Square . . . 76
    Skirmishers Individually resisting Horsemen . . 79
        Lancers, several Dragoons 82
            Examples 83
        Rendezvous 84

SECTION XIV.

Covering Movements—Observations 85
        Example 87
        Clearing Front 88
    A Battalion or Brigade 89
        Changes of Front 90
        Covering Troops Retiring 92

SECTION XV.

Light Troops in General Action 94
    Covering Troops Advancing to Attack 94
        Examples 95
    Covering Troops in Defensive Positions . . . 96
        Example 97
    Pursuing Defeated Enemy 100
    Covering a Retreat 101

SECTION XVI.

Covering Artillery 102
    Disposition 102
    Acting with Artillery in Out-post Affairs . . . 103
    Attacking Artillery 104
    Cavalry and Horse Artillery against Infantry . . 105

SECTION XVII.

Light Troops acting with Cavalry 108
    General Observations and Principles 109
    Disposition, according to Nature of Ground . . . 109
    Cavalry Dashing at Skirmishers 110
        Example 110
    Affording Mutual Support 111
        Acting on the Offensive 111
        Acting on the Defensive 112
        Acting Feeling for the Enemy 114
    French Intermix 114
        Chasseurs à Cheval 114
            Example 115

CONTENTS.

SECTION XVIII.

	Page
Advanced Guard—Light Division	117
General Purposes	117
On a Large Scale	118
Commander	119
Composition	120
Strength	121
Distance in Advance	122
Disposition on Line of March	122
Collateral Patroles	123
General Rule to threaten Enemy's Flanks	124
Precautions passing Defiles	124
Heights to be Crowned	125
Falling in with the Enemy	126
Small Advanced Guards	127
In Skirmishing Order:—	
Duty of Advanced Files and Flankers	130
Searching Villages, Defiles, &c.	131
Flanking Parties	133
Duty of Flank Patroles	134
Strength, Formation, &c.	134
Patroles of Discovery	136
Composition, Duties, &c.	136
An Advanced Guard pursuing the Enemy	137
Stratagems of the Enemy	138
Duties of Skirmishers	139
Examples	140—144
An Advanced Corps of Observation	145
Dispositions	146
Always to be held in Fighting Order	148
Example	149

SECTION XIX.

Rear Guard, General Purposes	150
Composition, Formation	150
Precautions before Retreating	151
Disposition, Stratagems to deceive the Enemy	152
Example	153
Covering a Retreat	153
Attention of Commander	155
How to check the Enemy	155
Examples	155—156
Rear Guard on Line of March	161

SECTION XX.

Attacking Bridges	163
Defending Ditto	164
Example	166

CONTENTS.

SECTION XXI.

	Page
Mountain Warfare	169
Attacking Defiles	170
Passage of Ditto	170
Defending Ditto	171
Retreating through Ditto	171
Forcing Defiles—Example	172
Khyber Pass, Affghanistan	172—177

SECTION XXII.

Defending and Attacking Woods	181
Attacking Ditto	184
Acting in Forests, such as America	186
Bush Fighting	186
Riflemen	186
Directions for Traversing Woods	187

SECTION XXIII.

Attack and Defence of Villages	189
General Observations:—	
Modes of Defending Ditto	190
Attacking Villages	193
Light Troops employed in Sieges	195
Covering Parties	195
Correctness of Fire required	195

SECTION XXIV.

Disembarkation of an Advanced Corps	197

APPENDIX.

General Rules for Skirmishing	201
Company Drill	213

INTRODUCTION.

ORIGIN OF LIGHT TROOPS.

LIGHT Troops have been more or less adopted by all Armies, whether ancient or modern—by barbarous as well as the most civilized nations. Their origin and utility would seem to have arisen from the necessity which large Armies were under (in order to ensure their safety) of sending out small parties to explore the country, to look out for the Enemy, to gain intelligence, or to collect supplies, &c., &c. As these parties would not have sufficient strength in close order to resist large bodies sent to oppose them, they would naturally have had recourse to strong ground, and availed themselves of shelter of every sort to protect themselves. Hence they would be obliged in some measure to separate in order to take advantage of trees, rocks, fences, &c., or such other cover as the ground offered in the Line of Defence; and thus their weakness, in many cases, might be turned to an advantage, because, by fighting under cover, their small numbers would be concealed, and they could shoot, whether from bows or with fire-arms, with greater effect, whilst the fire of larger bodies would be of little avail against Troops at open distances protected by cover; nor would the utility of this manner of fighting be confined to occasions of defence alone, but would be employed in favourable situations as a successful means of attack, since experience has proved that small bodies, properly disposed in dispersed order, have attacked and routed much superior force, as, for instance, the Indians in North America against General Braddock in 1757, and the Guerillas in Spain against the Veteran Troops of France. During the American War of Independence, and also at the commencement of the French Revolution, owing to the deficiency of well-organized and disciplined armies, the extensive employment of Light Troops was adopted; these proved themselves of the utmost utility, often vigorously attacking and worsting their enemies by swarms of Skirmishers alone, which has led to this description of Troops being more generally introduced and employed in increased numbers by all nations. Light Troops in the Grecian and Roman Armies derived their name from the nature of their arms, these being of a different and lighter description than those of the regular Troops; but since the invention of gunpowder, the difference has nearly disappeared. The true difference and the utility of employing Light Troops in the field does not now consist so much in their being differently armed (as some bodies are, such as Rifle Corps, Lancers, Cossacks, &c.) from Regular troops, but in their manner of fighting and the various purposes for which they are required in war.

The proper element of Light Troops is to fight in dispersed or skirmishing order, hence the system does not admit of precise forms and mechanical methods. The object is to defeat the Enemy, and prove their superiority over him by a skilful dexterity in the use of their arms; and in being good marksmen; and, as far as practicable, to fire, load, and move under Cover; in quickly seizing ground and taking advantage of every circumstance which the position or movements of the Enemy may offer; in fine, the faculty of quickly observing and judging for himself: courage and self-confidence, with bodily activity, are the peculiar attributes of the Light Soldier; so that this sort of fighting requires the abilities and personal energies of every individual engaged.

Frederick the Great said that Light Troops are, for the General, the torch which ought to enlighten him regarding the situation, the movements, and the nature of the designs of the Enemy.

The importance and utility of having good Light Troops is so indispensable that an army which has none, or only bad Skirmishers, would labour under great disadvantage in the present times.

Light Troops, organized on proper principles, and provided with qualified Officers, are of incalculable value to an Army, and are an invaluable instrument in the hands of a Commander who knows how to avail himself of their tact, activity, and self-confidence, whether to ensure the safety of his Army, to open the way for him to victory, or to enable him to obviate or avoid the evil consequences of untoward accidents.

They should be able to move rapidly from one point to another, to surprise the Enemy with audacious boldness; to see everything, as it were, without being seen because on the celerity of their marches and the concealment of their movements depend the success of their enterprizes.

As by the system of their formation Light Troops possess these qualifications, they are always employed in partisan warfare, which, according to General Valentine's definition, consists, "In all those military operations which have only for their object to aid and promote those of an Army or Corps, without having immediate relation to the conquest or defence of a country, and such ought only to be comprehended under the denomination of 'La Petit Guerre;'" hence, ensuring the safety of the principal Corps, and all those sorts of affairs which are calculated to annoy the enemy, constitute properly the object of Little War.

Light Troops, therefore, exercise great influence as essential auxiliaries in all movements, especially in the grand movements to which they are subordinate (to them are entrusted the advanced posts, reconnoitering, gaining intelligence, keeping up the communications, &c.), and if they do not fulfil well the desired object, the Army would be compromised.

The Light Service ought to attain the most perfect possible degree of perfection in every particular. It requires for the operations of war not only well-trained and experienced Troops, with skilful, distinguished, and indefatigable Officers, but also effective and determined Commanders. A General who would properly direct Light Troops in the field must possess clear judgment, tact, discernment, and perfect *coup d'œil*. He must know how, after winning one post, another ought to be seized; how to follow up a defeated Enemy with energy, but at the same time without rashness; or how to remedy the evil consequences of a disaster or defeat. He is frequently placed in trying situations: the duties are often so difficult that he requires the utmost energy and patient perseverance of his Troops to obtain a desired object. In continued bad weather,—in lengthened advances,—in arduous slow retreats, when want and hunger increase other manifold difficulties, the courage of the best Troops might be shaken, or they might become dispirited: besides, also, the constant vigilance, the frequent movements and restless activity which this service requires, is of such a sort that, in the long run, it may become irksome and oppressive. Hence it follows that Officers of Light Corps should be daring and enterprising,— men who possess the gift of knowing how to acquire a proper influence over their men, so as to teach them to overcome difficulties with courage, and to animate them to bear hardships with fortitude, both by word and example. Every Officer ought incessantly to study this branch of his profession in al its details. In order to attain this desirable object, and to improve himself, he ought to study history and narratives of war so as to impress the facts and spirit on his mind, and not merely load his memory with dry unprofitable rules.

In fine, intelligence, penetration, tact, address, with skill in stratagems, ought to characterize the Troops employed in this service, from the chief to the private soldier. It is essential that he who is destined to fight detached, independently or individually, ought to know how to conduct himself when in peculiar circumstances he may be abandoned to himself.

"An officer employed in partisan warfare has many opportunities of gaining experience, and of becoming a good Officer of advanced posts, so as to fill such a situation with success; because on many occasions he may have to act independently of the Army, and is often abandoned to his own resources. An Officer so circumstanced must unite address with courage, and boldness with discretion, in order to accomplish the objects of his mission without compromising his little Corps with a superior force. Being always surrounded by dangers, it is his business to foresee and surmount them. He thus acquires an experience in the details of war rarely to be obtained by Officers of the Line, the latter being always under the control and acting under the guidance of superior authority, which tends in a great degree to damp their energies, as it limits their experience, while the talents and genius of the partisan, having free scope, are developed and sustained by a dependence on his own resources."

Every Officer when going to service should provide himself with a good telescope, and a correct map of the country in which the Army is to act. These not only afford a great source of interest and amusement, but are essentially necessary to Officers on duty at the advanced posts. Indeed it is of the utmost importance that those employed in the van or rear-guards should, by a reference to their maps, make themselves acquainted with the roads, rivers, bridges, villages, towns, &c., and the general features of the country over which they are to advance or retreat. Telescopes for measuring distances, invented by Richaud, and made by Jones, Charing Cross, are valuable for such purposes.

SECTION I.

On the Utility and Employment of Light Troops.

GENERAL PURPOSES.
1. ACCORDING to the present system of warfare, the following seem the principal objects and reasons for employing Light Troops:—

2. When it is desirable to attain a proposed design by means of the smallest possible force;—in this view Light Troops are essentially required to ensure the safety of an Army in the field, whether it is in activity or repose; being as necessary to form advanced posts, &c., to protect camps and cantonments, as when they are engaged in covering the advance or retreat of Armies on the march, or the formations of Troops on the day of battle: also to act in detached parties in the various affairs of outpost, such as acting on the Enemy's Flank and Rear, surprising posts, intercepting convoys, collecting supplies, reconnoitring, &c., in fine, in all those diversified enterprises which constitute partisan warfare, and which require energy, concealment, and dispatch.

3. In very hilly countries, or thickly wooded and covered with forests, and in intricate ones having numerous passes and defiles; or in general, where the ground to be defended or attacked is so broken or intersected as to render the posting or movements of regular troops difficult, but which is at the same time favourable for the operations of Troops in loose order; therefore Light Troops are employed as the most efficient means of effectually defending, or of gaining possession of such strong ground. And hence it often enables a General in the field, even in general actions, to occupy an extent of broken ground with a small body of

Troops, and may so mislead the Enemy as to induce him to spend much ammunition very ineffectually.

4. Since the principles of strategy now adopted require that large masses of Heavy Troops should advance to bear on certain points, but never to be separated too far from their depôts and Reserves, and likewise to concentrate themselves in every new line of operations, a great number of Light Troops must be employed to establish and maintain the communications between the several Corps themselves, to unite them to their Reserves, and to reconnoitre the Enemy, and hold him in check.

5. The immediate application of Skirmishing on a large scale in the field is most frequently required on the following occasions:—

APPLICATION OF SKIRMISHING.

6. To protect the front of an Army by being pushed in advance as an Advanced Guard, so as to keep the Enemy at a distance from disturbing the main body. This comprehends all affairs of outpost which may occur.

7. In preparing or opening General Actions, and protecting the Lines or Columns, whether in position or advancing to attack, against desultory attacks, by Skirmishers being interposed to keep off the enemy, or to clear the way so that the Troops are not necessarily exposed to fire, until the decisive moment for attack or defence, while they may also endeavour to annoy the Enemy's Lines.

8. To conceal designed enterprises,—to mask an Attack by throwing out a swarm of Skirmishers to prevent the Enemy discovering our movements, and keep him in uncertainty of the true point of Attack, whether in the open field or in covering Troops advancing to storm a Fortress; likewise in executing false Attacks, or repelling those of the Enemy.

9. At the conclusion of battles after victory, to pursue and harass the Enemy, or to cover the Rear in case of defeat.

10. To protect the Rear of an Army as Rear Guard, when it retires in presence of an Enemy.

11. And, lastly, when it is necessary to engage when employed in all the various enterprises of partisan warfare.

SECTION II.

System of Training, and Qualifications required.

1. THE qualifications required of Light Troops, and the system of training them may be divided into two parts, viz., Mechanical and Intellectual.

2. The *Mechanical* consists in the accurate knowledge and dexterous use of fire-arms, with their effects, correct shooting in every position—judging distances, the different formations and combined movements of Troops in Skirmishing Order, the relative duties of the two men of the same File, the defence against Cavalry, whether in squares or individually, activity of body in getting over obstacles, the perfect knowledge of the Bugle Sounds, and a ready attention to conform to orders when under fire, so as to be able to rally quickly at all times, even when confusion may have arisen.

MECHANICAL.

3. The *Intellectual* comprehends tact,—adroitness,—acuteness,—"*coup d'œil*," and intelligence, so as to act with discernment under all circumstances, whether adverse or otherwise; a readiness of quickly taking advantage of cover, in every variety of situation, whether in open or broken ground; an enterprising disposition and energy to overcome difficulties; good discipline and daring intrepidity in attacking the enemy; ready to oppose him by force, or to deceive him by stratagem; by understanding how to deprive an opponent of every advantage.

INTELLECTUAL.

4. Therefore in forming Light Troops, care should be taken to select men endowed with proper mental and bodily qualities. Hence in regard to these, and the moral and physical powers required of them, particularly much continued perseverance,—the great trust

CHOICE OF MEN.

and confidence which must on many important occasions be reposed in them, the greater hardships, efforts, and difficulties they are exposed to,—Light Troops ought to be composed of better description of men than other Troops, who, fighting only in close bodies, require no personal energy of mind.

5. From the nature of the various duties expected of Light Troops, it is evidently of the utmost consequence to attend to the selection of men for this service as far as possible; so that only healthy, active, well-made men, possessing intelligence, and of sober, steady habits, should be chosen. Awkward, dull, stupid men, or those with weak chests, not having vigorous frames, or being short-sighted, &c., ought not to be taken.

6. A Skirmisher in the field is left in a great measure to act for himself. No mechanical forms guide his movements, which must be free without constraint. In all cases he must act with self-confidence, reflection, and judgment, according to the circumstances under which he fights. This requires intelligence and experience. Even the smallest intellectual ability and daring of a common soldier may tend, on certain occasions, to promote success. Light Troops being liable to be dispersed, and left to shift for themselves, much coolness and presence of mind are necessary under such circumstances. A few determined men, by showing a good countenance, may check a superior force. The success of Light Troops chiefly depends on the effect of each individual fire. A Skirmisher must know how to regulate his proceedings, that he may, by an effectual fire, occasion as much loss as possible to the enemy, while he himself, as far as circumstances will allow, is covered from the Enemy's fire. Therefore Light Troops, and Riflemen in particular, must possess a dexterity in the use of their arms, and prove their superiority in being expert Marksmen, so as never to lose a shot through awkwardness. They must never fire at random, but wait coolly for a fair opportunity, so as always to fire with effect. A soldier who has not sufficient presence of mind to reserve his fire for such opportunities, exclusive of wasting his ammunition, betrays a want of true courage, and throws a burden on his comrades, as he, without ammunition, becomes perfectly useless. To men acting in small parties, ammunition is extremely precious, and should be carefully husbanded. Noise and smoke are not sufficient

QUALIFICATIONS REQUIRED.

EXPERT MARKSMEN.

to arrest the advance of soldiers accustomed to war. They are only to be checked by seeing their comrades falling around them from the fire of a concealed Enemy.

7. Next to correct shooting, the skilful taking advantage of ground is the most important qualification of Light Troops. Every man must know readily how to take advantage of every circumstance of ground, according as the Position or Movements of the Enemy may permit, which can enable him to harass the Enemy without exposing himself, so as to deprive his opponent of every advantage; because this quick taking advantage of ground, and laying open the Position of the Enemy, essentially contributes to deciding Actions of Light Troops.

TAKING ADVANTAGE OF GROUND.

8. All these qualifications may be acquired in part by much practice. But a natural aptitude for the service will give the Skirmishers, who possess them, a superiority over those who, only after a tedious training, acquire a degree of usefulness. Instruction with practice is however always indispensably necessary. The men ought, therefore, to be thoroughly drilled in the close order movements, and fully instructed in all the formations, firings, and changes of position in extended order, according to Her Majesty's Regulations, and also to have acquired a knowledge of the Bugle Sounds previous to their proceeding to the practical application of Light Drill, and the various combinations required either for attack or defence, when on actual service in the field.

SECTION III.

Instructing.

1. IN all Instruction at drill or field practice, the attention of the men should be attracted by suitable explanatory remarks and practical lessons, and not merely by dry rules. They should be impressed with a just and complete idea of the things to be learned, as by this means superior and reflecting Skirmishers may be formed. To attain this end, and to draw forth the mental abilities of the soldier, the Officers and Non-commissioned Officers must use patience and perseverance. Neither time nor trouble should be spared. The Skirmishers, encouraged by proper Instructions and example, will soon learn to fulfil their obligations with zeal and pleasure.

<small>INSTRUCTING LIGHT TROOPS.</small>

2. They will be reminded that, although in Skirmishing the whole line act conjunctly, yet in each file a soldier acts individually in choosing proper cover, and in firing only when an opportunity offers. Therefore, that the uniformity and constrained positions so essential in movements in close order, with accurate dressing on all occasions, is here dispensed with, and that when Skirmishing in action they are permitted to carry their arms in the most convenient mode, so as always to be ready to shoot with facility. But in order to prevent their firing hastily and at random, when on actual service, they should be made, when at exercise, to go through the motions of loading and aiming with accuracy at some object, and to pause a moment after firing to watch whether his shot has taken effect or not.

<small>SKIRMISHERS ACT INDEPENDENTLY.</small>

Instructing. 7

3. In some situations they will have to conceal themselves by stooping or kneeling; in others by lying down, or creeping along the ground. They ought therefore to be able to load and fire with accuracy in every position.

FIRING IN DIFFERENT POSITIONS.

4. As the perfect proceedings in Skirmishing often depends on the judgment of single Skirmishers, so will the formation of an efficient Light Corps chiefly consist in the fundamental Instructions and practical exercise of the men together. This is best done at first by Company Drills, then two or more Companies united, and lastly, by the whole Battalion, which will enable every individual, whether Officer, Non-commissioned Officer, or Private, to become acquainted with, and quickly to comprehend (when acting on an extensive scale in broken ground) the particular and relative duties of the Skirmishers and Supports, &c. To further this object, when such Corps are at exercise, the Companies thrown out as Skirmishers should frequently be taken from the centre or other parts of the Battalion, and not always from the flanks, as is too generally the case. Nor should they always occupy the same position, but ought frequently to be chequered by being sometimes placed in the Centre or Flank, or in the Right or Left wing. And in order to accustom the men to act with intelligence on all emergencies which may occur on service, after a Battalion has been thoroughly drilled, there need be no hesitation in inverting the order of Companies in a Battalion, or Subdivision, or Section in a Company, whether acting in close or extended order; and when in the latter, they may be practised to disperse, and then to rally and form chain again in any required direction on two or three files placed for that purpose.

5. When Companies are Inverted, or if one or more Companies are detached from any part of the Line except the left; likewise when Companies who have been acting as Skirmishers are recalled, or suddenly driven in, and have formed on the nearest flank of a line, or in rear of a column, the whole will number off anew.

COMPANIES INVERTED.

6. In order to teach men the practical mode of readily Taking up Ground, they should be moved from one position to another across a country; and those that have not selected good cover should be corrected. This object will be much facilitated by posting a well-trained soldier in the same file with a young one.

TAKING UP GROUND.

7. Nothing gives a Soldier a clearer idea of what he ought to do in the various situations in Skirmishing, and teaches him better the quickly finding out and making use of trees, hedges, banks, ditches, &c., as cover, than making Companies oppose one another, each alternately acting in attack or defence. He thus readily acquires the system how he ought to act, according to the circumstances of ground and the movements of his adversary. See ample Instructions on these heads in Her Majesty's Regulations, pp. 317 to 320.

8. Skirmishers who do not possess the necessary adroitness and activity of body, not only cannot readily take the proper advantage of ground, but will be stopped by obstacles that occur, and which the more active enemy will avail himself of to his disadvantage. Hence the men should be practised in running, in jumping wide ditches, leaping over fences, &c., and also in swimming, so that no obstacle may deter them. These essential requisites may be attained by frequent practice and movements in a close intersected country, by which the Skirmisher in time will acquire a pliable activity, intelligence, and confidence.

MEN TO BE PRACTISED IN ATHLETIC EXERCISES.

9. Too many intricate and artificial formations should be avoided. They do not promote any additional flexibility in the troops, but rather occasion unsteadiness, and perplex the men, and might tend to disorder when engaged with the enemy, when all movements must depend on the circumstances of the moment, and be regulated accordingly.

INTRICATE FORMATIONS TO BE AVOIDED.

It is true that there are many Rules and Movements, both useful and necessary in the time of peace, for the better regulating Parade Field-day Movements, but which in action cannot be attended to, or literally put in practice before the Enemy.

10. With regard to Riflemen, whose principal object is to fight in ambush, a high degree of intelligence is required, they being on many occasions liable to be left to shift for themselves in detached Parties. They must know how to gain upon the Enemy, particularly if he is negligent in protecting his Flanks, in Advancing or Retreating. For these purposes they must be taught to steal along behind fences, walls, down ditches, through gardens, corn-fields, copses, passing quickly from one covered station to another, always taking care to load before advancing, or quitting the post where they fired last. They at the same time ought to act

RIFLEMEN.

with due caution, that they themselves do not get committed or cut off. In these enterprises they may move in file, or scatter or concentrate themselves, according to the nature of the ground or cover, and the object in view. But the two men of the same file at least must act in concert, and never separate.

11. A Rifleman ought to be well practised in the management of his Arms, which require to be in skilful hands. Every individual ought to possess a perfect acquaintance with his Rifle, and be able to judge of its effects, whether it carries true, and how to make allowance in firing to counteract any defect.

12. The Rifle is a most efficient arm when of a superior quality, and placed in very efficient hands. To excel in Rifle-shooting, a person requires to have a steady eye and hand, a ready capability of judging distances, with a knowledge of the influence which the state of the atmosphere and position of the sun have in estimating such distances; likewise a knowledge of the influence which the wind, and the conformation and nature of the ground over which the ball passes have on its flight and range; whether on a plain, or across a hollow, ravine, &c.; whether over marshy ground, or a small lake, or across a river, all which must be taken into account in correct shooting.

13. It is essential that Battalions of the Line should be practised in extended formations, so that, in REGIMENTS OF THE LINE. the absence of Light Troops, they may be competent to protect the Front and Flanks of a Column in march; *and hence* the formation of Advanced Guards, and the posting of Piquets, apply to all descriptions of Infantry. See Reg. Part V. Sect. 1, No. 4, page 254.

14. Indeed, according to the present system of warfare, all Infantry Corps ought to be so instructed as to be able to engage the Enemy either in close or extended order; because during the late Continental campaigns many of the General Actions principally consisted of attacks or defence on villages, entrenched houses, &c., or very rocky, broken ground, when the Troops were obliged to act often in detached bodies, or independently, and continually fighting alternately in close or extended order, as circumstances required, mutually supporting one another.

15. " Every Corps ought to strive to acquire the character of " a good Service Regiment. No higher compliment can pos-

"sibly be paid to any body of men than to be so styled; which
"implies its being famed for gallantry, discipline, and general
"good conduct in the field. Indeed this is the very point of
"excellence which every individual of a Corps who has its in-
"terests at heart must hope to see it attain. No Regiment
"ever proved itself a good and efficient Service one whose dis-
"cipline, *esprit de corps*, and general system and interior
"arrangements in every particular were not of the highest
"order previous to embarking for foreign service; and par-
"ticularly Light Corps, considering the variety and multiplicity
"of duties which such Corps will have to perform when called
"into the Field, and who are so liable to be employed on de-
"tached Service. Indeed, no Regiment will long continue to
"be effective in the field, when beset with the many hardships
"and vicissitudes inseparable from campaigning, that cannot
"boast of possessing all those desirable qualities of a good
"Service Regiment.

16. "It is not by merely entertaining military ideas of, and
"practising only no higher duties than Parade exercises.
"and Barrack and Stable duties alone, that Regiments can
"hope to arrive at any great pitch of excellence; or in the
"event of being called suddenly into active service, that they
"will prove themselves as well versed in all the most important
"duties necessarily required in the field when they find them-
"selves in close contact with an enterprising Enemy at the
"Out-posts." See Colonel Leach.

17. Hence an Officer in command, whether of a Regiment, Wing, Troop, Company, or other Detachment, cannot be better employed than by instructing those under him in the system of Attack and Defence.

18. If a small body, the Officer commanding it may send out a portion to occupy a favourable post as a Piquet. He may, with the remainder, move to attack the Piquet, preceded by an Advanced Guard and Flank Patroles, who will use the same precautions in exploring the ground over which they advance as would be observed during a Campaign in a hostile country. Much profitable practical information may by this means be acquired: if this can be done by a single Company, how much more by a Wing or whole Corps?

SECTION IV.

Duties of Officers and Non-commissioned Officers.

1. An Officer in command of Light Troops in the field
COMMANDING ought to possess great energy, decision, and
OFFICER. presence of mind. The greatest fault of a
Light Infantry Officer is the want of reflection
and self-confidence. The success of an Engagement, particularly in a woody or intricate country, depends as much upon the coolness and presence of mind of the Commander, the intelligence of the Officers, and the ready obedience of the men to orders, as upon their bravery.

2. He ought to be full of resources and prepared to meet every contingency, whether to take prompt advantage of fortunate incidents to insure victory, and to prevent defeat, or whether to remedy with ability any untoward accident or reverse of fortune. He should on all occasions endeavour to make himself acquainted with the ground on which he is to act; for this purpose he should ride forwards, or send on an intelligent Officer before hand to reconnoitre the country, so that on the arrival of the Troops they may at once occupy the most favourable positions, where, having ascertained the position or dispositions of the Enemy, he may readily proceed to the true points of attack; or if during a Retreat, he may be enabled to ascertain the next position proper for defence, or any Defile, Ravine, River, Bridge, Fords, &c., in the Rear, in order that the Line of Skirmishers may be directed to fall back in such direction, or be enabled to press such obstacles with greater facility.

3. On Service, many cases of emergency occur, and decided advantageous opportunities may offer, of taking advantage of errors the Enemy may commit, which justify a

deviation from orders previously received. Hence, on such occasions, all Officers must know how to act on their own judgment, without waiting for express orders, and particularly those in command of detached parties, or Patroles acting in the Flanks of Troops engaged.

4. The Commander will go wherever his presence may appear necessary, either for directing the whole or such particular portion, whether the chain of Skirmishers, the Supports or Reserve, as circumstances may require, according to the object in view, and the proceedings of the action. He must narrowly watch the motions of the Enemy, and avail himself of every opportunity that offers to obtain every advantage over the Adversary. Should there be a height close at hand, he may take post there to make his observations, and from whence he can better superintend the whole and direct the movements. If occupying a position or defending a Defile, &c., he will be careful to prevent his Flanks from being turned: for this purpose he will send out parties to protect them; and if acting on the offensive, previous to making an attack he will endeavour to turn the Flanks of the Enemy by detaching parties to move under cover of hills, rocks, ravines, wood, &c., with orders to dash on at a given signal, or when the attack in front proceeds.

He must invariably have a Bugler attached to him, and ought to employ a well-mounted Officer and one or two Non-commissioned Officers to convey his orders where necessary, or to report and communicate with the Officer in command of the Troops in his Rear.

5. Strict orders should be given, that men badly wounded are carried to the nearest Support, where the Officer will take down their names, and send fresh men to conduct them to the Reserve or Battalion, from whence they will be conveyed to the Field Hospital. By this means the Commander will know at any time the number he has lost, and it will obviate the inconvenience so general in war of seeing three or four men employed in carrying off one wounded man.

WOUNDED MEN.

6. When a Battalion is thrown into Skirmishing order, the Majors will in general have a place assigned to them by the Commanding Officer; the Senior one will always superintend that portion where the Commanding Officer is not present.

MAJORS.

Duty of Officers.

7. When Companies are employed in covering the movement of Troops in the Field, or those of a Battalion or Brigade, whether these consist of the same Corps or belong to different ones, a Field Officer is generally appointed to direct the movements. Should such not be the case, the Senior Officer on the spot will assume the command. He will be careful to conform to the movements of the Body he is covering, and that the relative positions and distances of the Skirmishers and Supports, with regard to such Body and between themselves, are duly preserved, as far as circumstances will permit. *See* further directions, Section XIV.

8. When a position is to be taken up, the Commander will, previous to the Skirmishers being extended, denote the points where the Flanks are to rest.

Captains.

9. Captains or Officers commanding Divisions should, previous to all operations, whether in the field or at exercise, appoint Officers and Non-commissioned Officers to each Sub-division and Section, which will prevent delay or confusion in the moment of any emergency.

CAPTAINS.

10. When a Company is ordered to Skirmish, the Captain will take post in its Rear, and give the necessary Orders, and will remain with the Sub-division in Support. But if the whole Company is to extend, he will in this case be in rear of the centre of his division of Skirmishers, and in all cases should be attended by a Bugler; and, on service, by a Corporal or steady man to carry orders.

Subalterns.

11. A Subaltern will be attached to each sub-division of Skirmishers to direct their movements. Should there be a third Subaltern, he will remain with the Captain to receive and convey his Orders. These Officers will superintend their Divisions with a watchful activity, going where their presence is necessary, and point out to the men the most advantageous cover. They must pay attention that the men preserve their order and distances from the point of

SUBALTERNS.

direction,—that they level and fire coolly, and load properly. They must observe the motions of the Enemy, and direct the men to the proper object of attack or defence. When casualties occur in action, the Officers will form complete files of such odd men as are left alone by the loss of their comrades.

12. The Officers in command of Supports are responsible for properly conducting them during the various movements for the choice of proper positions, particularly if required to form squares, or to take post and occupy defiles, &c., in retreating. They ought to keep the Supports as much under cover as circumstances will permit, so as to conceal them from view of the Enemy, and not present them as an object for his Artillery. But likewise in such situations that they may be ready promptly to reinforce or relieve the Skirmishers, when required. They must keep a good look-out that the Enemy does not pierce through any part of the Skirmishers, or that he does not turn or out-flank the Line. They are also to lead the Supports to make any attack with the Bayonet, should such be required.

OFFICERS COMMANDING SUPPORTS.

Non-commissioned Officers.

13. The Non-commissioned Officers will be distributed along the rear of the Skirmishers; and a due portion in retreating will retire with each Line. They will take care that the men attend to all the Instructions laid down for Skirmishing, correcting those that may require it. They will see that the men always get under cover when practicable, and that they never fire without taking a deliberate aim. They must be careful themselves to attend to, and to draw the attention of the men to, the Bugle sounds or Orders given by repeating or explaining them, and that they instantly cease firing on that signal. The Non-commissioned Officers ought never to fire themselves, except it is necessary in their own personal defence.

NON-COMMISSIONED OFFICERS.

14. As the centre is generally the point of direction, unless otherwise ordered, a Sergeant will always be placed in the centre of the Line of Skirmishers, whose duty is to direct the chain and lead the men to the proper point of attack, he must pay particular attention in superintending the directing File, especially in changes of Front, or when a new Line to a Flank may be formed.

15. It would be desirable if every Troop or Company when in actual service could be provided with a good Telescope of portable size for the use of the Non-commissioned Officers in command of Piquets and Patroles. Indeed during the Peninsular war the Officers were frequently obliged to lend their own glasses to their Sergeants and Corporals, when ordered on such detached duties on a particular service.

TELESCOPES.

SECTION V.

Ground, Positions, Cover.

1. GROUND may be considered under the following heads :—

2. Strong ground, such as Rocky Heights, Hills, Ravines, Dales, Woods, a close country intersected with Fences, Walls, Ditches, Canals, Farm-yards, Houses, &c.

3. Perfectly open ground, that is to say, open Plains, clear, smooth, undulating Heights.

4. Mixed ground, consisting partly of both the above, viz., either broken strong ground, having occasionally open plain places, or open ground, having only here and there Heights, Enclosures, Trees, &c.

5. Very broken ground is the most favourable for Skirmishers, particularly if not too close, or covered with underwood, so as to prevent their seeing the Enemy, or to render their movements difficult; as it admits of all the advantages that may be drawn from the proper use of ground. But in such ground, especially if wooded, Skirmishers must conduct themselves with circumspection, if the Enemy likewise occupy it.

6. Plain ground is the most disadvantageous for Light Troops. They must avoid open places as much as possible; and when this is not practicable, they must not remain longer there than is absolutely necessary. When obliged to engage in such places, they will lie down.

Position.

7. In throwing Light Troops into Skirmishing order the Commanding Officer will determine the Position to be occupied

according to the object in view, whether of attack or defence, and the nature of the ground. In choosing a proper position for the Line of Skirmishers, a Commander is not so limited in fixing on a defensive one as when he is to act on the offensive. The most advantageous position is such a one as offers to Skirmishers on one side the greatest number of objects to afford them cover, behind which they can fire conveniently, and which on the other side offers no opportunity or objects within range of their fire for the Enemy to cover themselves.

8. The borders of Woods, rows of Trees, Fences, crests of Heights, Rocky Ridges, Ditches, &c., are most eligible as defensive positions, particularly if in front of such places the ground is perfectly clear. In such positions the flanks of the chain should be made to rest upon some strong points, such as inaccessible Heights, Houses, Walls, &c. Where these are not to be obtained, a party should be detached to one or both flanks, as the case may require, to keep a look out, and prevent their being turned.

DEFENSIVE.

9. In offensive operations, when an attack is to be made against an Enemy already advantageously posted, the only choice of position consists in choosing a place where the Line of Skirmishers can be formed out of range of the Enemy's fire; and that the advance from this against him can be made with celerity. In such cases the Line ought to be prolonged sufficient to out-flank the Enemy; while parties should be sent to threaten and turn his Flanks.

OFFENSIVE.

10. Positions in general may be deemed to run from height to height; but in a country intersected with Fences, low Walls, &c., these form the first Position, and the nearest hedge, &c., across a field, will form the second one. In a Wood, the nearest row of trees will be considered the new Position, and so on in succession from one cover to another, whether in advancing or in retreating. When the movement in advancing consists of successive changes of Position, the Skirmishers, on attaining a favourable one, or when they have driven the Enemy from one, and have occupied it, will continue firing as long as circumstances may require, when they will proceed regularly to make another onset.

11. Retreating consists in taking up positions to the rear, or backwards, from the Enemy; that is to say, when cover admits, the Skirmishers will retire firing from tree to tree, &c. But when in giving up a Line of defence there is a plain or open ground, such as a

IN RETREATING.

large field between them and the next fence or cover, they will not defend such open space unless particular circumstances demand it, but will, with all celerity, proceed over it to gain such next cover. It may be considered a general rule in retreating, that while one part receive the Enemy, the other part ought to continue to retire into a new Position, to be repeated alternately as opportunities offer.

12. In taking up a Position or allignement, the Skirmishers in occupying it and posting themselves will conform to the nature of the ground, whether it is a serpentine fence or ditch, &c., having projecting angles, or is in a straight line. They will follow the bending of heights, rows of trees, &c., merely dividing the ground as equally as possible, without attending so minutely to dressing as to sacrifice any advantage of ground or cover. Should a knoll or a few trees lie a few paces in advance, which would admit of one or more Skirmishers laying down or resting their arms to fire, they will avail themselves of it. The Officers will, however, attend that the unity of the Line is not too much broken, and observe that no considerable gaps occur. They will assign other positions to such files who may be either too far apart or too much in front or rear. Care must be taken that Skirmishers never mask or get in the way of one another's fire. If any particular part of the ground is open, and should there not be cover either on one or both sides to admit of a few Skirmishers occupying such, and by a cross fire defend it, the Skirmishers must lie flat down and fire. Skirmishers must avoid getting into groups or crowding together in any part of the Position.

OCCUPYING POSITIONS.

Cover.

13. Skirmishers must recollect that although the position and general Line are taken up by the Commander, the men of each File are individually responsible for selecting proper cover. The chief advantages which Light Troops must endeavour to draw from the circumstances of ground are as follows:—

14. That Skirmishers, whether in position or in movements, are so covered as to be concealed from the Enemy, and greatly protected from his fire, and therefore enabled to direct their fire at leisure in some degree of security. That they can approach or steal upon the Enemy, to a proper shooting dis-

tance, or to change their position without being perceived. They must therefore learn to take advantage of every circumstance of ground to shelter themselves from the Enemy's fire, without by that either diminishing or hindering their own. Thus trees, clumps, fences, rocks, ruins, old walls, bushes, furrows, &c., are so many natural intrenchments of which they may avail themselves. They ought to comprehend the importance of concealing themselves from the view of the Enemy, and that they are when hid nearly in almost as great security behind a broom bush as behind a tree. Skirmishers when engaged either in advancing or retreating ought to examine the ground before moving, and then proceed without delay to the cover that appears most eligible. They must likewise know how to gain upon an Enemy's flank or his columns in march, particularly in advancing to attack, by stealing along behind fences, edges of woods, through cornfields, &c.

15. But although it is incumbent on Skirmishers to make the most of cover, it is not thereby implied that they are on that account to be confined to continued firing under shelter, when occasions offer to shorten the affairs by a decided attack with the bayonet. When opportunities occur to warrant such mode of proceeding, the whole Line will dash up together as will be hereafter explained in a future section.

SECTION VI.

Movements of a Line of Skirmishers.

1. THE movements of a Body in Skirmishing order may be divided into two parts.

2. The internal changes of the component parts of such Body, or the combined movements of Skirmishers and supports,—such as relieving or reinforcing the chain,—forming squares—new lines to a Flank, &c.

3. The general movements and changes of position of the whole, comprehending all operations of advancing to Attack, to cover a Retreat; to defend Positions, &c., and all enterprises in partisan warfare.

4. All Light Infantry movements in close order, when no particular time is specified, are made in " Quick Time." But all extensions and formations from file, and all closings or other formations from extended order will, in general, be done in " Double Time." And it will always be employed when Skirmishers are recalled by the "Assembly," or when Troops are withdrawn out of reach of the Enemy's fire. Too much haste in the field is disadvantageous. It gives rise to wavering, and leads to confusion. It aggravates the difficulty of directing an extended Line. It fatigues and exhausts the men, particularly when advancing to Attack, as it puts them out of breath, hence out of their power to Fire with precision. There are, however, particular occasions which require all possible expedition, such as when it is necessary to seize an advantageous Post, or when Parties are sent to out-flank the Enemy, &c. Likewise, ground which lies open to the effective Fire of an Enemy posted under cover, must be passed over as quickly as

TIME OF MOVEMENT.

possible. Such, for instance, as an open space in front of a Wood or a Fence, &c. And when Skirmishers are obliged to pass from one Position to another, where the men would be much exposed. Also in retreating from a Wood or other covered Position, if a clear Plain or a Field occur, such open tract must be crossed with all expedition, until another Cover is gained, or at least until they are out of reach of the Enemy's Fire, which he will keep up upon them from the strong ground they left, and which he will have got possession of. In such cases Running is permitted, and will be resorted to.

5. Every movement of a Line of Skirmishers may be considered as a Change or Passage from one Position to a new one, which will be successively changed for another as often as the object of the operation makes it necessary.

6. All the movements of a chain should be without constraint. Yet in order to preserve the unity of the Line, and that every movement may be properly conducted, especially when acting in a wood or across an intersected country, POINT OF DIRECTION. some one point within itself must be fixed upon for the purpose of regulating and guiding the whole as a pivot on which each movement hinges, as well as a point of union. Therefore a particular file, according to the object in view, will always be named as the "Point of Direction," whether a Centre, a Flank, or any other file, as the case may require,—from which distance and dressing is to be preserved as far as circumstances will permit. But too much attention to Dressing, even at Exercise, cramps DRESSING. the habits of the men in judging for themselves in selecting cover, which is the principal point; because the Skirmishers are supposed to execute their individual movements, whether in Advancing or Retreating, according to reflection and the advantages of Cover. Each File must, therefore, observe the proper moment to change from one cover to another, without depending too much on those on their right and left. They will, however, preserve as long as possible the same order as they held in Close UNITY OF THE LINE. Order. To attain this, and to prevent Breaches, they must occasionally correct Dressing and Distance from the point of direction, which also is the principal means by which the Line can be directed, and it will enable them to rally or reform again in Close Order when required. But in continued movements, and in the course of prolonged action, from diversity of cover, casual and other

circumstances, this is not always practicable. It is, however, essential that the two men of the same File never separate, and that in a wood, high corn, &c., each File ought to keep sight of the next File nearest them towards the point of direction, so as to prevent their being too far separated or any dangerous gaps occurring. They must likewise avoid crowding too close together, or ever collecting in groups at any particular points, as by doing so they would present a large object to the Enemy, and draw a concentrated Fire upon themselves, and thus defeat the very intention of fighting in skirmishing order. In extensive woods, very broken ground, &c., or in Attacking villages where the view is interrupted, and the whole Line cannot be directed by the Commander, the course of the Action and the ground, as it varies, must determine the place of the Skirmishers, when the unity of the Line cannot be entirely preserved. In such cases the Officers commanding each Division must guide it according to the passing events, and will endeavour to act so that their operations may be in combination with the others, and tend to forward the object in view. No looseness must be permitted. This would prevent orders from being properly obeyed, and render every movement—rallying or assembling—almost impossible. Hence at all times the utmost exertions must be made to maintain the unity of the Line, and to recover it if lost.

NOT TO COLLECT IN GROUPS.

7. When Skirmishers are in movement, whatever the Directing Point may be, whether a Centre or a Flank File, on their being ordered to take ground to a Flank, either direct or oblique, the leading File becomes the directing one. Also if a Wing is thrown forwards or backwards, the Pivot File is the point of direction. And in either case, when the Line resumes its parallel position, the former point of direction must be taken up unless another is named.

PIVOT.

8. As the Halt annuls all previous sounds except the "Fire," Skirmishers will, therefore, on the "Halt" sounding, instantly do so; no matter during whatever manœuvre they may be executing. If marching to a Flank, they will turn to the proper front. If Retreating, they will face about so as always to remain Fronted towards the Enemy when Halted. And if they were firing previously, they will continue to keep it up.

HALT.

9. It may be held as a general principle in the movements of Light Troops, that all Advances and Formations to the Front should be from the Centre as much as possible, and that all retreats should be from one or both Flanks.

GENERAL RULE.

SECTION VII.

SKIRMISHING ORDER.

Dispositions to Engage.

1. In throwing Light Troops into Skirmishing Order, there are certain general principles which must be adhered to with regard to the disposition to engage; the number to be extended, the manner this is executed, and the direction of the Line, &c.

2. It must be held as a general rule on all occasions that the whole body destined to Skirmish ought never all to be brought into action at the same time; and that whatever strength the force employed may be, never more than one-third, or at most one-half, should be thrown forward to engage as Skirmishers. The remaining part will be formed into such separate bodies of Supports, with or without a reserve, as the nature of the particular service may require, for the following purposes, viz., to support the Skirmishers by reinforcing or relieving them in case of their being pressed, and as points for the Skirmishers to rally upon, or to form Squares to resist Cavalry, &c.; and when a part is held in reserve, a compact body is always ready to act on emergencies, easily moved to any point most threatened or pressed, or to follow up any advantage the Skirmishers may gain, &c.

<small>GENERAL RULE.</small>

3. The disposition and formation of Light Troops to engage depend on the object in view, and may be reduced to two heads. First, when a small force is to act as Skirmishers, merely to cover the movement of Regular Troops; and Second, when Light Troops are detached in a large body to act independently of the main Corps or Army.

Dispositions to Engage.

4. Under the First Head, when Skirmishers are only thrown out to cover the Front and mask the formations, or to protect the movements of Regular Troops, equal Parties in support will be sufficient; and a Reserve will not be necessary, as the Lines or Columns which they cover, and on whom they depend, answer the purpose. In this case, when the chain of Skirmishers and their Supports are considered only as a cover to protect or prepare the way for the movements of Troops, during operations either of Attack or Defence, &c., the movements of Skirmishers will depend upon, and must be regulated according to, the object in view, by the manœuvres executed by the Battalion or Brigade in their rear, to which they will conform on all occasions.

SKIRMISHING ORDER.

SKIRMISHERS AND SUPPORTS.

5. Under the Second Head, when a body of Light Troops is detached from the main Body to act independently in a particular service, or to keep the Enemy at a distance, such as Advanced or Rear Guards, and all other affairs of out-posts, where a serious attack is intended and resistance to be encountered, a more formidable disposition, and more regular formation becomes necessary. The Troops on such service, when about to engage, must be disposed so as to support and cover each other in three separate bodies, at due distance in rear of one another, viz., as Skirmishers, Supports, and a Reserve; that is to say, the Line extended as Skirmishers must not only have Parties of Support in the Rear, but also a strong body in Reserve in rear of the whole, ready to move up and act as circumstances may require. In this case the Supports and Reserve will conform to, and their movements will in general be regulated by, those of the Skirmishers in their front, as it must be recollected that the movements of the Skirmishers will depend upon, and will be guided or determined by, the dispositions or resistance of the Enemy.

SKIRMISHERS.

SUPPORTS AND RESERVE.

6. When Light Troops are thrown out to Skirmish, the distances to be taken by the chain and Supports will be regulated according to circumstances, such as the nature of the country, whether intersected or open; whether the Enemy's Skirmishers are supported by Artillery or Cavalry, &c. When no particular distance is ordered, the Supports will be ad-

26   *Dispositions to Engage.*

DISTANCE. vanced 150 to 200 paces in front of the Line or the Reserve, and the chain of Skirmishers extended 100 to 150 paces in front of the Supports.

7. These regular Divisions of Light Troops may on some occasions suffer certain modifications. Thus Supports may not be required when Skirmishers are only thrown out at a short distance to cover temporarily a Regimental formation. And, on the other hand, when a Line of Skirmishers is of great extent, and separate Attacks made by single Companies, in particular points, such as in attacking a village, or field-works, &c. It is the practice of some Continental Armies, that, exclusive of the Supports and Main Reserve, each Company, in such cases, especially the Flank ones, should leave a portion as a Local Reserve, which they term " Replis."

8. The duty of the Skirmishers is to engage those of the Enemy,—to drive them in,—or, if acting in the defensive, to keep them at a distance by a well-directed fire, &c. The Supports will be ready promptly to render such aid to the Chain as circumstances may require. Their number will depend in general on the length of the Line of Skirmishers, and their strength will be regulated by that of the force employed. The general rule is, that they should be composed of equal numbers to the Skirmishers. The Reserve is destined to give effectual support to the bodies in front, and to act against the Enemy when nothing more can be performed by the Skirmishers, whether from their being driven in, or that it is wished to follow up some decided advantage which they may have obtained over him.

9. The particular duties of each of these bodies, with their relative duties to one another, will be detailed hereafter.

10. Skirmishing and all Out-post duties are better performed by whole Companies than by portions of several.

UNITY OF COMPANIES. When a Light Corps is employed to Skirmish, the unity of the Companies should be preserved as much as possible. The Skirmishers and Supports may be, when on a small scale, composed of the same Companies; but it is more desirable, when acting in the Field that the Skirmishers and the Supports should each be composed of separate and entire Companies. By this means each one is more easily united, or, if attacked by Cavalry, can more readily throw themselves into Squares of Companies, or, if there is time, into Squares of Grand Divisions, by the

Skirmishers retiring and forming on the Companies in support. Moreover, the men are urged to a zealous discharge of their duties by a chivalrous spirit of emulation between different Companies. Comrades are also brought in active duty together: a mutual competition and confidence are thus established, which are of much importance when engaged in dangerous service.

SECTION VIII.

Mode of Throwing Light Troops into Skirmishing Order.

1. In throwing a body into Skirmishing Order, the strength of the Enemy and the nature of the country, with regard to the advantage of cover it affords, will determine how much of the force should be employed as Skirmishers, and how much should be formed as Supports. The more open and plain the ground is, the greater must be the number of Supports. More Skirmishers ought never to be thrown out than the strength displayed by the Enemy, and the extent of the ground to be attacked or defended, absolutely requires; because, if the Line of Skirmishers was overcrowded, men would be sacrificed without necessity, from all of them not being able to find cover; nor would they have room to load and fire conveniently, so as to give a well-directed fire. More men than necessary would tend to create confusion, and there would be less disposable strength left in reserve for the decisive moment of attack or defence, or other contingencies. Therefore, the Fire Line should at first consist of as few people as the case may warrant: it can afterwards be increased as circumstances may require. A few, but skilful and practised Skirmishers, in advantageous ground, soon make an Enemy give way, although more numerous, if less expert. Hence precise rules suitable to all cases cannot be laid down regarding what number of Skirmishers are to be extended,—the manner in which this is to be executed,—or the position of the Line. These depend on the nature of the ground, on the strength and position of the enemy, and whether the intention is to cover an advance, or a retreat. Therefore, when Light Troops are thrown into Skirmishing Order, the manner of doing so, and the formation

STRENGTH EMPLOYED.

of the Chain of Skirmishers, whether by extending them on the spot or while on the March, advancing or retreating, or whether from the Flank or Centre of any named Division, will depend on the above circumstances, and the object in view, as each particular case may require.

2. If the object is merely to cover the Parade movements of a Battalion or Brigade, or on actual service, when the Formation is made at a considerable distance from the Enemy, and the ground is favourable, the extension is usually done on the March, as directed in Her Majesty's Regulations, Part V., Sect. 3, No. 14. But in a difficult country, and in order to ensure concealment, the Troops may be led in Close Order up to the position, and then extended from the Halt along the line of attack or defence, as directed in Her Majesty's Regulations, Part V., Sect. 3, No. 1, leaving the Supports at due distance in the rear. And in this case, where more than one Company is employed, each Division may be marched up to that point of the position where it is to act, and each extend in succession when the regulating Company has finished; while those that are to form the Supports will at the same time proceed to occupy their respective positions.

EXTENDING ON THE MARCH.

HALTED.

3. The distance which a Chain of Skirmishers should be in advance of the Supports, and the latter in front of a Battalion or Reserve, must depend on circumstances, and the cover the ground affords. In general, the Supports may be about 150 paces from the Column or Line, and the Chain of Skirmishers about the same distance in front of the Supports. In a plain, or where there is any apprehension of Cavalry, the distance ought not to be too great; but on strong ground of difficult access to horsemen, they may be at a greater distance, particularly if the Enemy have Artillery.

DISTANCE.

4. Previous to all formations in Skirmishing Order, the Commander will notify what portions will form the Chain, and what the Supports. He will likewise denote the point from which the extension is to commence; whether from a Flank, the Centre, or any named File of a particular division.

5. The usual distance between each File of the Chain is six paces, when no other is named. But there can be no positive rule for the exact distance: this must depend upon the length of line which a certain number of Skirmishers are required to cover; that is to say

BETWEEN FILES.

the distance between the Files must be in proportion to the ground to be occupied, and which the men must divide as equally as possible. In such cases, the Commanding Officer will, previous to the Skirmishers moving out, denote to the Officers of the Centre and Flank Divisions the extent to be occupied, so that they may regulate the proportional distances to be taken according to the length of the position. In woods, the trees, or in broken ground, the nature of the various kinds of cover will determine the particular distance between the Files; because although the Commanding Officer determines the general position and direction of the Line, yet each File will select its own particular cover, whether the distance may be a little more or less on one hand or the other, or a little backwards or forwards.

6. When the extension is completed, the Centre becomes the point of direction, if no other is notified; and on service, the Skirmishers, after having secured cover, will wait for further orders, whether to fire, advance, or retreat, &c.

7. But in covering parade movements of a Battalion or Brigade at Field Exercise, when Skirmishers are thrown out to cover the immediate advance of a Line or Columns, they will, after extending, continue to move on until further Orders are sounded.

8. However, if they are only to cover a deployment, they (unless otherwise directed) will, in general, after extending, occupy the best cover at due distance in front, and wait for further orders, whether to fire or advance, &c.

Companies thrown out from a Battalion.

9. When one or more Companies are thrown out in Skirmishing Order, to cover the movements of a Battalion, each leaving a Sub-division in Support, the following are the usual modes of executing the Formation—

From Line.

10. When a Battalion is in Line, if a single Company is ordered to Skirmish, the Commanding Officer will notify which Sub-division is to remain in Support. On the sound "To Skirmish," the Sub-division to Skirmish will advance, extending from its Right, Left, or Centre, according as it may be a Flank or

ONE COMPANY

Centre Company; while the one to form the Support will proceed in quick time, to occupy its proper position in rear of the Centre of the Line of Skirmishers, inclining when necessary to whichever hand may be required.

11. If the Right and Left Flank Companies of a Line are to Skirmish on the sound "To Skirmish," after moving out a few paces, the Outward Sub-divisions of each will continue advancing in double time, extending towards the Centre; while the Inward Sub-divisions will move on in quick time to take Post at due distance, in Rear of the Outward Flanks of the Chain.

TWO COMPANIES.

12. If the Light Company be posted by Sub-divisions on the Flanks of a Battalion, it will proceed in same manner, each Section acting as above directed for Sub-divisions.

LIGHT COMPANY.

13. Should two Companies from each Flank be ordered, the Flank or outward ones will advance and extend; and the two next or inward ones will proceed to form the Supports, as above directed for Sub-divisions.

FOUR COMPANIES.

14. But if a Company is only sent out for the temporary purpose of Covering a Manœuvre, a Support may not be required, in which case the entire Company will move out and extend, as above; but will not advance to so great a distance.

15. If the two Flank Sub-divisions only are to cover the Movement, they will advance extending from their Outward Flanks towards the Centre.

16. In all those cases, after the extension is completed, Dressing and Distance will be taken up from the Centre unless otherwise directed.

17. In the above examples, the Outward Flank, Sub-divisions, or Companies, have been thrown out as Skirmishers, but the Inward ones may be extended and the Outward ones left as Supports, at the discretion of the Commanding Officer.

From Column.

18. When a Battalion in a close or quarter-distance Column, if the front Company be ordered to Skirmish, it will advance leaving such Sub-division in Support as may be named. The other will extend from the Centre, or from either the Right or Left Flank as may

ONE COMPANY.

be directed, according as it may be intended that the Column is to advance or retire, or whether it is to be deployed on a Central or a Flank Division; while the one left in Support will in each case take post at due distance in Rear of the Centre of the Line of Skirmishers.

19. When two Companies are ordered to cover a Column, or one which is about to deploy on a Central Division, and are immediately to extend while advancing, they may move out to each Flank to clear the front, either by Sections of Threes, or Filing, or the Diagonal March, according as the Column may be at close or quarter distance, or as the nature of the ground admits. And when clear, the Inward Sub-divisions will front, each extending from their Inward Flank, while advancing. Their Outward Sub-divisions will continue to gain ground obliquely outwards, and take Post as Supports in Rear of the Outer Flank of the Chain.

TWO COMPANIES.

20. But if the two Front Companies of a Column are ordered to extend to a Flank, for instance to cover a Battalion right in front, which is about to deploy to the left, the front Company will advance, leaving its Outward Sub-division in Support, and extending its Inward one to the Left. The Second Company will at the same time Face and follow in the same direction, and its Inward Sub-division will front and commence extending when the Front Company has finished. While the Outward Sub-division will continue to proceed to take Post as Support in Rear of its Outward Flank.

21. If the Column is to deploy to the Right on the Rear Division, the first Company will face and proceed to the Right, either in Sections of Threes or by Filing. The next Company, when clear, will advance, leaving the Outward Sub-division in Support, and the Inward one will extend to the Right. When this is accomplished, the Inward Sub-division of the First Company will front, and take up the extension to the Right, while the Outward one will continue to proceed to take Post as Support in Rear of the Right Flank of the Chain.

22. If Three Companies are to cover a Column, on the sound "To Skirmish," No. 1 will lead out to the Right; No. 2 to the Left; No. 3, when uncovered, will advance, extending a Sub-division from its Centre, leaving one in Support. When No. 3 is

THREE COMPANIES.

finished, the Inward Sub-divisions of No. 1 and No. 2 will front and extend to complete the Chain, while their Outward Sub-divisions will continue to gain ground to either Flank to form the Flank Supports.

23. In these cases the Inward Sub-divisions are extended; as those being nearest the Centre, less ground is required to be gone over, and thus time saved.

24. If a front Company is sent out to cover without a Sub-division in Support, the Company will advance extending from the Right or Left Flank, or its Centre, as it may be ordered, according to the object in view.

SINGLE COMPANY.

25. Companies ordered from the Centre or Rear of a Column will face towards their respective Flanks, File out, and, when clear, will Front, Turn, and proceed as directed for Front Companies; while those which are to form the Supports will, at the same time, Face and move out to occupy their proper positions in Rear of the Chain of Skirmishers.

COMPANIES ORDERED FROM THE CENTRE OR REAR OF A COLUMN.

26. In these examples the Companies have been thrown into Skirmishing Order while in movement; but in an intricate country, or where a covered position is to be occupied, the Troops for such Service may take Post about 150 paces in front of the Body they are to cover, where the Supports will remain; and then the Companies which are to form the Chain may move on in Close Order, taking distance at the same time from the regulating one, until they arrive at such Posts of the position as it may be judged that each will occupy; and will then take up the extension in succession from the regulating one, while those left in Support will proceed to Right or Left, as may be required, to occupy their proper positions.

27. When the intention is to mask a Retreat, the portion destined to form the Chain may advance a few paces in front of the Body they are to cover, and will there extend on their own ground, so as to hold the position and protect the movement, while the Supports (and Reserve when there is one) will proceed to the Rear of the Main Body, each halting and fronting when at their proper relative positions and distances. When the Lines or Columns have retired and gained a sufficient distance to the Rear, "the Retreat" will be sounded, on which

TO MASK A RETREAT.

the Troops in Skirmishing Order will commence retiring, and the Chain will fire or not, as may be directed.

28. When Light Troops are posted on the Flanks of a Brigade, whether such is in Line or in contiguous Columns, they will, when ordered to cover, move to the front and extend to Right or Left, as may be required. Or if certain portions are held in Rear of a Division, they will move through the intervals of Battalions, and extend in such direction as may be required, always proceeding on the same principles, and leaving equal numbers in Support.

29. When two or more entire Companies are to form the Chain, having also an equal number of Companies in Support, or if the Battalion is to be thrown into Skirmishing Order, they may proceed as directed in the two following Sections.

30. Skirmishers and their Supports will conform to the movements of the Battalion or Brigade in their Rear. For instance, if the Battalion advances, they will push on; if it retreats, they will follow; if it moves in Echellon, they will proceed by the Diagonal March; if it takes ground to a Flank, they will Face and move in File in the same direction; if it changes front to a Flank, they will form a new Line to that Flank, &c. In fine, if the Skirmishers are not called in while the Battalion performs any movement, or changes its formation, they must, with the utmost rapidity, also change their situation so as to correspond with the new order of the Battalion, and protect it during the change. But further details are given in Section XIV.

CONFORM TO MOVEMENTS OF BATTALION.

31. If one Company be covering, and if it be not recalled when the Battalion forms Square, it will clear the front, and will form Rallying Square on whichever Flank offers the best position; but if recalled, it will retire by the shortest way to the Rear, and form the Rear face of the Square, having taken care to open to Right and Left so as never to run across the front of a Column or Square.

WHEN A BATTALION FORMS SQUARE.

32. If two Companies are covering, and if they are not recalled, the Skirmishers will retire, each Wing falling back on their respective Supports, and will form Square with them.

33. When the formations of a Battalion are completed, it is essential that the front should be cleared as soon as possible. Skirmishers when at Field Exercise are usually recalled in the following different manners:—

CLEARING THE FRONT.

34. If the Skirmishers are to assemble on the Line they hold, a Point will be denoted by a distinguishing " G " followed by the " Close," on which they will face to the Point indicated, whether Right, Left, or Centre, and close in upon it, and await further orders.

35. But when the " Close " is sounded without a " G," the Skirmishers will retire, each division closing on its own Centre on the March; and will form on their respective Supports in the first place; and the second " Close " or " Assembly " will be the Signal for the whole to rejoin the Battalion or other Body,

CLOSE.

36. When the " Assembly" sounds first without any "Close," it is the signal for both Skirmishers and Supports to make the best of their way to the Rear; for this purpose, if covering a Battalion, they will separate from the Centre and move rapidly to each Flank, if it is Line, or to its Rear if in Column.

ASSEMBLY.

If covering a Brigade, the Centre of the Chain may retire through the Intervals of Battalions, and those on the Outward Flanks by the Flanks of the Brigade.

37. The Supports in these cases will fall back in steady double time.

38. The examples given above for throwing Companies out to Skirmish from a Battalion may be requisite to ensure uniformity at Exercise or Field-day Practice; but all Companies of Light Troops must be prepared to act promptly in every contingency which may occur. They are liable in the Field to be suddenly called upon from any part of the Battalion, and to be employed upon diversified services. In such cases, whether one or more Companies or a whole Wing is required, either to act singly or in an united command, the Senior Officer of each such detached party will observe that the Troops are thrown out in proper Skirmishing Order. The same rules for doing so (particularly for retaining a portion in Reserve), hold good from a single Company to the largest Body. On such occasions, when two or more Companies are employed, whether to drive in the Enemy's Piquets, to occupy a Pass, to hold a Post, or any other particular service, the due proportion of Skirmishers, Supports and Reserve, will be regulated according to the number of Companies, always observing that it is better that each of these Bodies should be formed of entire Companies when circumstances admit.

REMARKS.

36 *Mode of Throwing Light Troops*

39. Indeed on service, Light Corps generally adopt the excellent plan of taking all tours of Out-post Duty, Picquet &c., or other detached commands by Companies, so that Comrades always act together under their own Officers.

40. During the Peninsular War, the greatest part of the 5th Battalion, 60th Royal Rifles, were attached by Companies to Brigades in different Divisions of the Army, and were found eminently useful on many important occasions, as may be gathered from the Duke of Wellington's General Orders, strongly recommending them to the favourable notice of the General Officers.

A whole Battalion to Skirmish.

41. When a whole Battalion is thrown into Skirmishing Order, either to cover the movements of a Brigade, or to protect the formations of Troops about to engage without any portion of it being left in Reserve, the Corps for this duty will take post at about 150 paces, or such other distance as may be named, in front of the Body it is to cover, if at Exercise; but on Service, the distance will be determined by the nature of the ground; and the formation may be executed either from Line or Column.

42. If the Battalion is standing in Line, the following simple rule, whether it consists of eight or ten Companies, seems the readiest mode of executing the movement.

FROM LINE.

43. The two Centre Companies, and the two Flank ones, will form the Supports; all the others will advance to skirmish, a central one of these being named as the regulating one.

44. Suppose the Battalion to consist of eight Companies, on the sound "To Skirmish," Nos. 2, 3, 6, and 7 move out in double time. No. 3, the Regulating Company, instantly commences extending while advancing from such named File as will be sufficient to cover the interval left by the two Centre Companies, while No. 6 will at the same time extend from its Inward Flank. Nos. 2 and 7 move on obliquely Outwards, and as soon as Nos. 3 and 6 have finished, they will each "Front, Turn, and Extend." The Companies left in support will proceed as follows:—The two Flank ones, Nos. 1 and 8, will face Outwards, and proceed to take post in rear of each Flank of the Line of Skirmishers. The two

Centre Companies will remain on their own ground, and will be formed into Column of Sub-divisions by the Senior Officer.

45. If the Battalion consists of six Companies, the two Flank and one Centre Company may form the Supports, while the others move out to form the Chain, viz., No. 3 will extend from its Centre on the March; No. 2 and 5 move on obliquing to Right and Left, and commence extending; when No. 3 has finished, Nos. 1 and 6 at the same time proceed Outwards to form the Flank Supports, and No. 4 will remain as the Centre one.

46. Or it may be done by alternate Companies: suppose the Right ones are to remain in support, the Left ones will move out; No. 4 will extend from its Centre; Nos. 2 and 6 Oblique Outwards, and extend. When No. 4 has finished, No. 3 remains as the centre Support; Nos. 1 and 5 proceed to form the Flank ones.

47. If the Battalion is in Close or Quarter-distance Column, TO DEPLOY AND EXTEND. it may deploy and extend according to the following dispositions, whether it consists of eight or ten divisions. If of eight divisions, on the sound "To Skirmish," Nos. 1 and 2 Face and lead out to the Right, No. 4 to the Left. No. 3, when uncovered, advances, and when parallel with No. 2 (which will have halted and fronted at six paces from the Column), both will move forward, extending from their inward Flanks. Nos. 1 and 4 will continue gaining ground obliquely outwards until the Centre Companies have finished, when they will "Front, Turn, and Extend." Simultaneous with the above, Nos. 5 and 6 will Face Outwards, and proceed to form the Flank Supports, while Nos. 7 and 8 will remain on their own ground as the Centre one.

48. If the Battalion consists of only six Companies, it will deploy and extend as follows:—No. 1 will face and proceed to the Right, No. 3 to the Left. No. 2, when uncovered, will advance, extending from its Centre, and when it has finished, Nos. 1 and 3, who had continued obliquing Outwards, will Front, Turn, and Extend; while at the same time No. 4, having faced to the Right, and No. 6 to the Left, will proceed to form the Flank Supports; and No. 5 will remain halted on the Centre one. The Battalion will thus stand by wings in two Lines, the one as Skirmishers, and the other as Supports.

49. These formations should likewise be practised with the Columns left in front.

50. In all the above formations, the Companies which are to form the chain of Skirmishers, instead of extending while advancing, may, *when it is deemed necessary*, be marched up in close order to the position they are to occupy, and then extended.

51. The Officer commanding the Centre Support will always keep in front of the centre of the Brigade in his rear, and will conform to its movements in conjunction with those of the Skirmishers; while the Flank Supports, by watching the movements of the Centre one, will always conform and move parallel with it.

52. When a Battalion in this order is threatened by Cavalry, if the Skirmishers have not time to fall back and form squares with their Supports, as directed in Section XIII., each wing of the chain may close to its own centre, and form squares; or if much pressed, each Company may in the first instance form Rallying Squares, and then unite in larger ones, as circumstances may admit. The Supports in these cases will form squares on their own ground, ready to move up to assist the others.

TO RESIST CAVALRY.

53. When the Skirmishers are ordered to clear the front without forming on their Supports, they will retire through the intervals of Battalions; but those on the Outward Flanks may run round the Flank of the Brigade.

54. If the Battalion is to be re-formed on its own ground, the Skirmishers will retire, each Company closing on the march to the flank nearest the Central Support, and then Inwards turn and form upon it. The Flank Supports will rapidly close in at the same time, and the Battalion will be re-formed in Column or Line, according as they find the Centre Supports formed, by the Covering Sergeants taking up the covering upon it, as the case may be.

BATTALION RE-FORMED.

Light Infantry Order.—Skirmishers, Supports, and a Reserve.

55. When a body of Light Troops is detached to act on a particular service, such as an Advanced Guard, Rear Guard, or other affair of Outpost, whatever the strength of the body may be, the general principle must be observed, that, in addi-

tion to the portion thrown into Skirmishing order, a due proportion must be held in reserve.

56. When a Battalion, whether consisting of eight or ten Companies, is ordered for such duty, it may be formed in Quarter-distance Column. The two front Companies will extend as Skirmishers; the two next will form the Supports, and all the others will remain halted in reserve. Thus suppose eight Companies are in Column, on the sound " To Skirmish," No. 1 will instantly advance, extending from its Left; No. 2 will move on, inclining to the left sufficient to clear six paces beyond the Flank of No. 1, and then extend from its right; at the same time Nos. 3 and 4 will face Outwards, and proceed, gaining ground obliquely to Front and to their respective Flanks to form the Supports; the others remain as reserve.

FROM COLUMN.

57. If the Column consists of six Divisions, the two or three front ones, as may be deemed necessary, will advance to Skirmish, each leaving a Sub-division in support, as directed ; the others will form the Reserve.

58. When the formation is to take place from Line, the two Flank Companies will advance, extending from their outward flanks ; the two next will move out to form the Supports; the others remain in reserve, and may either continue in Line or be formed into Column at quarter-distance as circumstances may require. Should they remain in Line and have to pass a bridge or defile advancing, it may be done in double Column of Sub-divisions or Sections, or by double Files from the centre, and after passing the front, increased by re-forming double Column.

FROM LINE.

59. When Skirmishing in this order, due care should be taken to prevent the flanks of the chain from being turned. For this purpose parties may be thrown out on one or both flanks, as may be required. But on ordinary occasions a double File thrown out from the Supports in a diagonal direction beyond the flanks of the Skirmishers to keep a look out, may be sufficient; should a stronger one be necessary, a Sub-division or Company will be sent from the reserve.

60. The movements of a body in this order will depend upon those made by the Skirmishers, to whom the Supports and Reserve will conform, because the movements of the Chain must depend upon the dispositions and resistance of the Enemy.

61. If the Skirmishers advance, the Supports and Column in reserve will follow, each preserving their relative distances.

62. If the Skirmishers advance, or retire a wing by right or left shoulders forward, or change front to the right or left, or form a new Line to a flank, the Column will conform, and will proceed to occupy its place in rear of the centre of the new front or Line of Skirmishers by the diagonal march and wheeling.

<small>SUPPORTS AND RESERVE CONFORM TO SKIRMISHERS.</small>

63. If the Skirmishers retreat, the Column will face about and retire to the nearest favourable position for defence; and if the Skirmishers are driven in by superior numbers, it will there deploy, to protect them and repel the attack.

64. If the Skirmishers are unable to dislodge the Enemy from a position, and that it becomes necessary to make an attack in close order, the Column will be deployed, and the Skirmishers recalled.

65. If an attack of Cavalry be apprehended, the Skirmishers and Supports will form squares and act as directed, and the Reserve will form square on its own ground, ready to move up to protect the retreat or union of the smaller ones.

66. When Skirmishers are recalled by the "Close," they will proceed as directed for that sound.

67. But if the "Assembly" be sounded, and the Battalion is to be re-formed in Column, the Covering Sergeants of Companies will run back and take up covering on the reserve, while the Skirmishers and Supports will retire, each Company of the former closing on the march to its inward flank, and every one, when opposite its proper place in the Column, will file into it.

<small>BATTALION RE-FORMED.</small>

68. If the Reserve be in "Line" when the Battalion is re-formed, the Skirmishers and Supports when re-called will make the best of their way to the respective flanks to which they belong; and each Company will form in succession four paces in rear of the formed flank, viz., a Sergeant of No. 2 will place himself four paces in rear of the right flank of the formed Line, and a Sergeant of No. 1 will place himself four paces in rear of the right of No. 2; while the same will be done by the Companies on the left flank: then each Company in succession will be dressed up into the Line in the usual manner.

69. When the Chain is to consist of more than two Com-

panies, and that three or more are thrown out to skirmish, the formation will take place as directed by Her Majesty's Regulations, Sec. 4, Part V., No. 2, p. 278.

70. When a Brigade, whether consisting of two or more Battalions, is employed on a detached service, and if a very extensive chain of Skirmishers should be required, one Battalion may form the line of Skirmishers and the Supports, by being thrown into Skirmishing order, as detailed in Sec. VIII, No. 43, and the other Battalions form the Reserve.

SECTION IX.

Skirmishers Firing in Action.

1. SKIRMISHERS ought to be made to understand that the object when brought into action with the Enemy is not merely to occupy certain positions and fire at one another, but that of acting on the offensive. When they proceed to attack, the design of their engaging is to assail and overcome the Enemy; they must therefore assail him with utmost vigour, and, if acting on the defensive, their duty is to resist and vigorously repel his Light Troops, and to retard the advance of his heavy masses by every possible means which the nature of the ground will admit of.

2. When Skirmishers are to engage, the Commanding Officer will, in the first instance, give the order to fire according to the distance of the Enemy or other circumstances. But although the sound to " Fire " gives permission to do so, it however remains at the discretion of each Skirmisher at what moment or how often he is to do so, because no Skirmisher ought to fire until an Enemy is within effectual range, and that he has covered an object with a probability of hitting, as it may happen that a part of the Enemy's Line is within range and another not, or some of his Skirmishers more exposed than others. Skirmishers therefore ought to be accustomed to begin to fire or not, only as opportunities may offer; otherwise numbers would throw away their fire, and ammunition not only be uselessly expended, but an enterprising Enemy, seeing a whole line unloaded at the same moment, might make a dash up before it was again prepared to resist; hence Skirmishers

(marginal note: EACH SKIRMISHER ONLY TO FIRE AS OPPORTUNITIES OFFER.)

must never fire at random, but each will select an object and take deliberate aim, and by correct firing, prove their superiority over the Enemy. Their eyes should be continually directed to watch and discover any exposure of the Enemy, while they themselves are concealed. They ought to fire not so much at those directly in front, as at those a little to right or left, as these, being more exposed, they may take them in Flank. "They should never let slip the opportunity "of firing on the moment an Enemy passes from one cover to "another. The first object should be to fire at the Enemy's "Officers, and to direct their fire to where they may hear Bugle "sounds or commands given." * They ought never to fire at

RANGE TO FIRE AT. a greater distance than 200 paces, except against a Column or Artillery, when they may fire at 300 paces. As a general rule, if they fire at 300 paces, they ought to aim at the head, if at 200 paces, at the shoulders, if at 100, at the middle.

3. When in action, each file of the Chain will fire independently as opportunities offer, without any reference to those on their right or left; but both men of the same file should never fire at the same time. When one has fired, the other must not do so until he observes whether his comrade's shot has taken effect or not, nor until his comrade has reloaded, so that they may never both be unloaded together, but that one will always be ready to fire, and cover the other whilst he loads. This seconding secures them from being taken by surprise, and inspires them with confidence.

4. There are, however, some occasions where this alternate firing need not be so strictly adhered to; for instance, in strong defensive positions behind walls or entrenchments, where they are well secured from the Enemy's fire, it is evident each Skirmisher should fire and load without delay. In such situations, or in defending narrow defiles, &c., the best marksmen may be selected to fire, and the others to load.

5. On all occasions, each Skirmisher, the moment he has fired, will instantly load in the mode most convenient to himself; when firing lying on the ground, the best position to load is to turn on the left side, prime, and then place the butt of his piece against his left foot, or between both, and load leisurely, without raising his right elbow too high.

* Prussian Regulation.

To turn on the back or from one side to another to load requires a greater cover than 24 inches high.

6. The position of each Skirmisher in firing, whether standing, kneeling, or lying, must depend upon circumstances, and the nature of the cover each occupies. Each ought to adopt that position by which he can best fire with effect and without consraint.

FIRING IN DIFFERENT POSITIONS.

To fire standing ought never to take place except from behind trees, walls, corners of houses, &c. To fire standing in open ground is incompatible, and contrary to all rules of skirmishing, if it can be avoided. When obliged to engage in such exposed places, the men ought to lie down, taking advantage of furrows, dung-heaps, large stones, &c., or place the chaco, with a handkerchief, tobacco-case, or grass, &c., put into it before them. And if separated from the Enemy by a river, or other impassable object, where they cannot be suddenly dashed upon, they may place their knapsacks before them as a guard. Should the ground be marshy or too wet to admit of their lying down, they must keep moving from side to side so as to prevent their adversaries having a fixed mark to fire at. To fire kneeling is in general the most convenient position, particularly when low cover is at hand, such as stumps of trees, &c.

STANDING.

7. The two men of the same file must always remain together and act in concert, particularly in advancing or retreating, so as never to separate, not only for the purpose of mutual protection, but because this will facilitate rallying quickly, forming squares, &c., or in defending themselves against single horsemen.

TO ACT IN CONCERT.

8. Skirmishers must on all occasions instantly stop firing, the moment "Cease Firing" is sounded. Likewise, when any Rallying—Assembly—Form square,—or a dash up and attack with the bayonet is to take place. If they are to stop firing by order, the nearest will stop, to be taken up by those on the right and left, and the word passed on.

CEASE FIRING.

9. Skirmishers in the field generally carry their arms at the trail; but in damp weather it ought to be carried under the right arm to preserve the lock dry. In quick movements, climbing hills, leaping fences, &c., they may carry it in the manner most convenient to themselves.

10. On actual service, Skirmishers will in general only fire when halted; and although Light Troops should be taught to fire and load when either advancing or retreating, yet when engaged in the field, if they are properly commanded, it will not often be required that they should fire when in movement because should there be, in advancing, an open space between them and the enemy, it ought to be dashed over as quickly as possible, to the nearest cover or inequality of ground, and there halt and fire, ready to push on again to the next advantageous position. The same system will be adopted in retreating; on abandoning strong ground, the Skirmishers after firing there will withdraw, and proceed with all celerity over any open space to the nearest favourable position; then halt and open their fire and maintain themselves until the enemy presses closely on, when they will proceed as before. Hence there are few cases when it is proper for a line of Skirmishers to fire while in movement; it in general ought only to take place against Troops retiring in confusion, or Column of the Enemy which have neglected to cover their rear and flanks. It may likewise be necessary when a Column of ours retires, to protect it, and keep the enemy at a distance from annoying it; or sometimes, when a Column, or contiguous Columns advance, to attack, having Skirmishers on the flanks and intervals.

FIRING HALTED, OR IN MOVEMENT.

11. In cases where the enemy in certain positions keeps up a galling fire, and is not to be shaken or readily driven off by our fire, the whole line of Skirmishers may simultaneously make a dash up with fixed bayonets, and carry his line of defence, should not an evidently too great a superiority of numbers, or the inaccessible nature of his position prohibit such a measure. Also in advancing, feeling for the Enemy, on approaching within 60 or 80 yards of the edge of a wood, a fence, or ridge of a height, &c., when there is reason to apprehend such place may be occupied by the enemy, the Line may in like manner dash up, fully prepared for attack or defence, in case the enemy may be lying concealed; and on all such occasions detached parties must be sent to threaten or turn the Enemy's flanks.

SIMULTANEOUS ATTACKS.

12. Example.—When the Left Wing of the British, under Sir Thomas Graham (now Lord Lynedoch), forced

the passage of the Bidassoa and established themselves in France, the defile of Puerta de Vera had to be forced previous to the advance of the right wing. To ensure the success of this operation, the Light Division were ordered to drive the enemy from the heights above the town of Vera, supported by the Fourth Division.

"The 1st Battalion of the Rifle Brigade was destined to "commence the attack by driving the enemy from a high and "rugged hill which flanked the right of the pass, which was a "necessary operation previous to the defile itself being "attacked. Colonel Ross threw the Battalion into skirmishing "order and extended the line of Skirmishers, so as to "encircle its base, and thus out-flanked and enveloped "the flank of the enemy. The whole chain then pushed "right up the hill without almost firing a shot, notwithstanding "the sturdy resistance of the enemy, and in a short time "completely cleared this formidable height. The French "Commander, finding himself so daringly pressed in front, "and his flanks threatened, closed all his force—formed line "in close order—and fixed bayonets, apparently with a deter-"mination to charge our Skirmishers; but on their reaching "the summit, the enemy went to the right about and "retreated. This dashing operation was the admiration of "the Fourth Division.

"Shortly afterwards, the two Passes of the Puerta de Vera "were carried after a formidable resistance."*

Firing Halted.

13. When Skirmishers fire "halted" they are always supposed to be under some sort of cover. Each one may fire in such a position as each particular circumstance of cover may require. On these occasions and in all cases of firing, good judgment and practice in actual service will direct and guide Skirmishers better than can be taught by theory.

FIRING HALTED.

14. On ordinary occasions, when firing halted, whether standing, kneeling, &c., the Skirmishers will proceed as directed, Reg. Part V., Sec. 3., No. 4, 6, 7. Each front-rank man will fire when an opportunity offers, and when he has loaded; the rear-rank man will look out for an

* Surtees, Twenty-five Years in the Rifle Brigade, p. 240.

object and fire, and so continue. Should a piece miss fire, or the one that fired not hit an enemy advancing on them, his comrade will immediately take aim and fire as soon as the other has reloaded.

15. In firing from behind single objects, a Skirmisher will always, when practicable, point to the right of it, and after firing, step to the left and rear to admit of his comrade occupying it, and covering him whilst he loads. Thus, in a wood, he will always have a tree on his left side as a guard, if he finds a branch at a convenient height, he can lay his rifle over it as a rest. Should a tree or other object not be sufficiently thick to cover both men, each will take separate trees as contiguous as possible to one another; but if each man cannot get a tree, the rear-rank man ought to lie down at the foot of his comrade and fire in that position.

IN COVER.

Firing Advancing.

16. In offensive operations when Light Troops proceed to attack in skirmishing order, the great object in advancing is to drive back the Enemy's Skirmishers rapidly and in confusion on their Reserves, giving them no time for rallying or making a stand. On these occasions, in the field, Skirmishers advance in a general line, pushing on or falling back from post to post, or from one covered position to another, so as never to stand exposed. Therefore when obliged to advance across an open space upon an enemy posted under cover, a quick and simultaneous dash towards such point should be made; a regular systematic advance over such open ground would entail a great and useless sacrifice of lives. Therefore, if the country is of such a nature that the advance is conducted by a quick passage from one position to another, as for instance, across fields from one fence to another, &c., the Skirmishers will only fire when they have attained each new position or cover.

IN GENERAL LINE.

17. But when engaged and working their way, advancing in very broken ground or cover, particularly in a wood, both men of the same file must be loaded before pushing on, and the one who is ready to fire will always be in front. A Skirmisher ought never to leave cover until his comrade

WORKING THEIR WAY THROUGH COVER.

is loaded, nor advance ever so little without being covered by the other having his piece ready to fire. Therefore, in such cases the man in front will steal quickly on to gain the next tree, &c., while his comrade will protect his advance by aiming at the enemy. As soon as he has secured cover, the other will close up to him, where both may continue firing alternately as many shots as circumstances and the general movements of the line may sanction; they will thus continue to act in mutual concert, supporting one another, and creep from one cover to another, and never directly expose themselves if possible. They will endeavour, by an occasional glance of the eye, to keep distance from and to avoid getting too far before or behind the file next them towards the point of direction.

18. If the advance be over ground affording little cover, and that the Skirmishers must move firing in line, on the sound to "Fire" the front-rank man of each file fires, then steps to the left, and commences loading on the march, following the rear-rank man, who will have continued advancing slowly with his piece in a threatening attitude; as soon as the other has closed up, he will give the word "Ready" to the comrade in front, who will then fire and load, &c., as directed, and so alternately continue. (Reg. Part V., Sec. 4, No. 5.)

19. But on service, it is often difficult for the men in broken ground to load on the march, especially Riflemen; it retards their movement;—much powder is lost,—and it prevents their loading properly; it is therefore preferable in these cases that they should halt to load, while they will afterwards regain by an accelerated pace the time employed in loading; or they may at all events, after firing, stop to prime and put in the cartridge, then move on, finishing their loading, and run up to their file leader. (Reg. Part V., Sec. 3, No. 10.)

LOADING.

20. The French, when it is required to make a rapid advance over a plain, or ground which is so open as to offer no cover for either party, or position of defence, and clear the way for their Columns and Cavalry, by driving in the Skirmishers of the Enemy, form their chain of Skirmishers three deep, and the men of each file are numbered 1, 2, 3. On the sound to "Advance and Fire," No. 1 of each file runs forward ten paces, fires and loads. The others continue to follow in quick time until

FRENCH MODE.

they reach No. 1, when No. 2 dashes out ten paces, fires and loads, followed by Nos. 1 and 3. When these join No. 2, No. 3 advances, fires, &c., and so they continue without intermission.

Firing Retreating.

21. In retreating before the Enemy, it is essential to retire with great coolness, and always to have a portion ready to fire, in order to impose upon him, to keep him in check, and also to be ready to repel a sudden dash.

22. If the Retreat is from cover to cover, the line of Skirmishers will only fire from behind the position they have gained. When on leaving cover they have to pass over open ground, they must, with all expedition, gain the next one, or, at all events, get beyond the accurate range of musketry before coming to a stand, thereby rendering the shelter they are leaving useless to the Enemy, and will oblige him in coming on to pass over such space exposed to a destructive fire which they will open upon him, and when he presses closely on, they will again retire as before. Reg. p. 273, No. 8.

PASSING OPEN GROUND.

23. During a Retreat, each Skirmisher after firing ought to scan the ground he is to traverse in falling back; and having selected a cover, he will gain it with all expedition. Thus, when Skirmishers retire through a wood or broken ground (which is best done in General Line), on the sound to "Fire," the man next the Enemy will fire and immediately run to the next best cover or tree in the rear, and load; while the other will remain kneeling, and protect him by keeping his piece pointed towards the Enemy, upon whom he will keep his eye until his comrade is ready. He will then fire and retire, passing beyond the one who is halted, to the next cover, and so continue alternately. But if the Enemy does not press on very close, he will, as soon as his comrade is loaded, follow without firing, occasionally looking back towards the Enemy, and when an opportunity offers, he will front, fire, and proceed as before, carefully observing that the man who is loaded always remains in the rear next the Enemy.

IN COVER.

24. When Skirmishers are retiring in a General Line, or coming to a stone fence, wide ditch, &c., the Rank in rear will remain on the defensive on the side next the Enemy until the other has got over and is prepared to fire, when the rear one will clear it with the least possible delay.

25. It is always difficult to effect a retreat with regularity in the presence of an Enemy. Advantage must be taken of every favourable circumstance of ground to arrest him and prevent his pressing too closely. In such case the best mode of conducting a retreat so as to resist the Enemy with advantage and give relief to the Skirmishers, is by alternate lines of Skirmishers and Supports,—viz., when in retreating the Supports arrive at a favourable position for defence, they will occupy it, and extend as Skirmishers, while the old Skirmishers, on reaching the new line, will pass rapidly through it, and form up into Supports. When the retrograde movement continues, they will again in their turn occupy as Skirmishers the next best cover that may offer. This method gives greater confidence to the Troops, and obviates any confusion or discouragement which might arise if Skirmishers, pressed by the Enemy, should make a mistake in not occupying a position properly. But by this means the Light Troops coming out of fire being formed into Supports, have time to rest, and to put their fire-arms into order again, and complete their ammunition. Should, however, one part be much exhausted, an equal portion from the Reserve may occupy the next position as Skirmishers so as to relieve them.

ALTERNATE LINES.

Alternate Ranks.

26. In ordinary Parade Movements, and on some occasions in the field, the Skirmishers may retire by alternate " Ranks," on the sound " Fire and Retreat." If the Line is halted and kneeling, the front-rank men fire, and move to the rear, loading on the March. When loaded, they will halt front, and kneel down in the position of Making Ready. The rear-rank men will then fire and retire in Double Time, passing to the proper left of the front-rank men; when they will take up the Quick Step and commence loading, and when finished, will front and kneel down, &c. The front-rank men again fire, &c. They thus continue to retire alternately, as soon as they hear the ramrods working of the rank that retired. Reg. p. 265, No. 11.

27. In action, however, when cover presents itself, the men will always avail themselves of it, and may load after halting and fronting under cover. In this case, the distance to be taken by each rank will depend on the move-

Skirmishers Firing in Action.

ments of the Enemy, whether he presses on rapidly or not, and the nature of the ground and the cover it affords, as at one time such may be near, and at another further off.

28. On all occasions during a Retreat, should the "Halt" be sounded, the line of Skirmishers with their Supports will immediately front towards the Enemy, each file getting under cover; and if they had been firing, they will still continue to do so.

29. The French on some occasions, particularly when pressed on ground not affording good cover to make a stand, retire in a line of Skirmishers formed three deep; each file is numbered by ranks 1, 2, and 3. When they are to engage, No. 1 of each file halts and fires, and then follows the others who had continued retiring. As soon as No. 1 joins them, No. 2 halts and fires, while the others move on, and when No. 2 joins, No. 3 halts, fires, and joins. The No. 1 again, and so on, all loading in turn on the march, so that there is always one firing, one ready, and one loading.

FRENCH MODE.

Firing taking Ground to a Flank.

30. When a line of Skirmishers takes ground either obliquely or direct to a flank, keeping up their fire, the leading file is the directing point. Each file will steal on from one covered place to another, preserving intervals by each taking up in succession the position which the Leading Files had left, and thus each file relieves the preceding one.

31. When a line of Skirmishers are to take ground to a flank, firing on the March, the whole will be faced to the one required, and will move in extended files in that direction. If to the left, on the sound to "Fire," each front-rank man will halt, face towards the Enemy, and fire. The rear-rank man moves on obliquely sufficiently to his right to get on the Line of Defence. When at about twelve paces more or less, as cover may offer, he will face towards the Enemy; and when his comrade who had followed loading on the march reaches him, he will fire, while the front-rank man will in turn proceed on, &c., and so continue alternately.

32. By these means the two men of the same file are kept together, and will never be both unloaded at the same time. Likewise each separate file of the chain will occupy, as they advance, the place or cover vacated by the preceding one, so

that they preserve their intervals by relieving one another in succession as they progressively gain ground to the flank. The Supports also face and move in the same direction.

33. If the "Halt" be sounded, the whole halt front towards the Enemy, kneel down, and continue firing. If the "Advance," they front, turn, and proceed as directed for firing in advancing. If the "Retreat," they likewise front and proceed as directed for firing and retiring. If "Cease Firing," they instantly cease; Files get into their places, and continue gaining ground to the Flank.

SECTION X.

Duty of Supports.

1. IN accordance with the general principle that when Light Troops are thrown into Skirmishing order a portion must remain in Support, no line of Skirmishers ought ever to engage without having sufficient Supports in its rear, on whom the chain may depend for succour in case of need. The duty of Supports is various and of an important nature, requiring great vigilance and attention on the part of the officers. The following are the principal purposes for which they are required:—

2. In the event of the Skirmishers being overpowered by superior numbers, or threatened by Cavalry, that they may have a point to rally upon or to form Square.

UTILITY.

3. To reinforce and strengthen by filling up gaps of any part of the line that may be pressed or forced.

4. To relieve the whole or any part of the Line that may be required.

5. To cover and protect the flanks of the Line from being out-flanked or turned, so as to prevent Flank attacks of the Enemy.

6. To out-flank the Enemy by moving up to prolong the Line to one or both Flanks as may be required, or to make a Flank Attack.

7. To charge in close order defiles, bridges, barricades, or roads, &c.

8. With regard to the strength of the Supports, as they are destined to relieve the chain of Skirmishers when required, they ought to consist of equal numbers, that is to say, there will be as many Companies or Sub-divisions in Support as the Skirmishers are composed of;

STRENGTH.

so that in general each particular division of the chain will have its own Support, with which it will be in communication, and on whom it will depend for assistance, and which will, unless otherwise directed, remain in rear of its centre.

POSITION OF. Yet in some cases the distribution and position of the Supports must be determined by circumstances and the nature of the ground. For instance, should one part of a line of Skirmishers occupy very strong ground, while the other part is exposed to danger on a plain or open space, the latter will require more Supports, and these to be closer to one-another than the former. When an obstinate resistance is expected, the Supports may be stronger than the usual proportion. But there is danger to increase the number of Supports by subdividing them into small parties; because, should any part of the chain be overpowered, these too small bodies, instead of being able to make a stand, would probably be borne off along with the Skirmishers in confusion; nor would they have sufficient union or strength to resist an attack of Cavalry. The number of

NUMBER. Supports must depend upon the length of the line of Skirmishers; but whatever the number of Supports may be, there must always be one in rear of each flank of the chain. In a very extensive line, having numerous Supports, the distance between them from one to another should never exceed 200 paces. In woods or broken ground, where heights, fences, &c., may prevent the Supports from seeing one another, a Corporal or a trusty man should be stationed to keep up the communication and to observe the Enemy, and give notice should he force any part of the chain.

9. In skirmishing across a country either in advancing or retreating, one of the chief duties of Supports is to protect the Skirmishers from a sudden dash of the Enemy's Cavalry. They will be ever ready to present themselves wherever the chain is most exposed to danger from the Enemy's horse, or at such parts as appear most vulnerable, and not merely follow at random behind the line of Skirmishers; and at all times will act with circumspection, keeping a watchful look out to front and flanks that these may not be turned.

10. A case in point occurred during the affair on the Coa, 24th July, 1810. Two companies of Rifles were warmly engaged, retiring before a far superior force, while the wing of another Corps was posted in support along a low stone wall about 100 yards in the rear; when suddenly a

squadron of French Cavalry, which had got round and turned their left flank, penetrated between the Skirmishers and their Support, dashed upon the chain, trampled down and sabred the men, who, being gallantly occupied with the Enemy in front, were thus taken by surprise; hence they could offer little or no resistance, so that most of them were killed or taken, but few escaped. It would appear that the Support had not taken the precaution to have thrown out parties or patroles in the flanks, and on observing the Cavalry, mistook them for some of the Germans in our service, were thus thrown off their guard, and allowed them to penetrate before they discovered the mistake.

11. The distance of the Supports in rear of the Skirmishers will be regulated according to circumstances, such as the proximity of the Enemy or nature of the country, whether the weather is clear or hazy, or whether it is day or night. In a close country where the chain is exposed to be out-flanked, the distance should not exceed from 100 to 150, and in no case above 200 yards.

DISTANCE FROM SKIRMISHERS.

12. Whether the Troops are in position or in movement, great attention should be paid to the choice of ground where Supports are to be placed. The position selected, when practicable, should conceal them from the view of the Enemy, and cover them from his fire and artillery so that they may not be unnecessarily exposed, but which will at the same time admit not only of their maintaining a free communication with the Reserves in their rear, but also will permit of their moving with facility to give a prompt and effectual support to the Skirmishers when pressed, or should the Enemy penetrate any part of the Line.

COVER.

13. Supports will generally advance or retire in Line, but if an attack of Cavalry is apprehended, they will move in Close Column of Sections, or Sub-divisions, according to its strength. In taking ground to a flank, it will be done in Column of Sections, or by Filing, according as the nature of the ground may admit. The flank Supports, when moving either in Sections or by Filing, will regulate as each may require to have their Right or Left in front, so as to have the Front Rank facing outwards in case they should suddenly be required to prolong the Line or repel a Flank Attack.

IN MOVEMENT.

14. The movements of the Supports depend upon, and must

be in combination with those of the Skirmishers, so as always to maintain their relative situations. Therefore when a line of Skirmishers charges front or position, the Supports will conform to the movement; and to ensure accuracy, one Support (as the case requires) will be named as the directing one, by which the others will regulate their movements.

CONFORM TO SKIRMISHERS.

15. Supports will " Trail " arms, when ordered to move, and " Order " them when halted without any word of command.

16. During an Advance the Supports and Reserve must not be too eager to press forward, but will give time to the Skirmishers to feel their way, by stepping short or stepping out, as may be required to preserve their relative distances.

17. When a line of Skirmishers changes front by Wheeling, or throws a Wing backward or forward, by Right or Left Shoulders forward, the Supports will immediately conform to the movement, and proceed rapidly to occupy their respective place in the new position in rear of the Line. Or when Skirmishers take ground to a Flank, the Supports will face to the same, and move parallel with them. In fine, whether Skirmishers are acting in the field or are covering the formations of a Battalion, or large body at exercise, the Supports will be prompt to act in combination with them.

18. When Skirmishers are recalled by the "Close," the Supports will remain until the Skirmishers join them, &c., as directed, Section VIII. But when Skirmishers are recalled by Order of the "Assembly," the Supports will not wait for them, but will instantly retire in Double Time, on the Battalion or Reserve, and get into their place in Column or Line, as the case may be.

RECALLED.

19. On the smallest apprehension of Cavalry, the Supports ought to be particularly on the alert. Each will form Close Column of Sections and advance, looking out for good positions, and prepare to form Square. On the "Alarm" being sounded, they will endeavour as much as possible to interpose between the Skirmishers and the Cavalry so as to give them time either to fall back and form Square conjunctly with them, or to form Rallying Squares on their own ground. If the Skirmishers do retire on them, the Squares will be formed by Sections as directed in Her Majesty's Regulations, Part II., Sec. XXIII. p. 65, or by Sub-divisions according to their united strength, the Skirmishers forming the Rear Face.

RECEIVING CAVALRY.

20. But should the Attack be so sudden as to oblige the Skirmishers to form Rallying Squares, the Supports will in this case form separate Squares, and will use every endeavour to aid those formed by the Skirmishers.

21. On these occasions the Supports should advance steadily in Quick Time. Too much haste would exhaust the men, and prevent their firing, when necessary, coolly or with effect.

22. Should the Enemy's Cavalry turn a Flank, and attempt to penetrate, the nearest Support will move in Column of Sections towards them, and endeavour to keep them in check, either by forming Square or occupying any advantageous cover for defence that may be at hand, until the Skirmishers can form Rallying Squares.

23. When Squares formed of Skirmishers and Supports are
SQUARES REDUCED. to be reduced, and the chain to be re-formed, on the sound to "Skirmish," the former Supports will dash out and extend, as being a fresh body; while the former Skirmishers will remain as Supports, and will complete their ammunition, clean their arms, and load.

24. When the "Retreat" is sounded, the Supports and Reserve go to the right about and retire; and the Skirmishers, if they are to fire, remain for the moment fronted to the Enemy, and then commence firing and retiring.

25. The duty of Supports during a Retreat is of a very
RETREATING. arduous nature. Should the Skirmishers be much pressed, they must give them such effectual Support as will check the pursuit of the Enemy, either by reinforcing them, or, if driven in, by making a stand to cover their Retreat, or, by taking up good Position in the Rear, ready to extend and relieve them; which is the best mode, because a Retreat is always best conducted by alternate Lines. Therefore the Supports when so ordered, during a Retreat, upon arriving at favourable Positions of defence, as very broken ground, fences, edges of woods, entrance of a defile, &c., will occupy them, and extend, while the old Skirmishers will retire through them, and form the Supports, and so continue acting alternately. On this and on all other occasions, when Supports are to relieve Skirmishers, they will act as directed, Section XII.

26. The Supports will likewise, during a Retreat, on coming to a ravine, a river, marsh, or other obstacle, endeavour

to ascertain the best passages of bridges, fords, &c., which they will secure by taking post there, and sending notice by a trusty man to the Skirmishers to inform them of the best Lines of Retreat, that they may withdraw in the proper direction.

27. When the Skirmishers are engaged with the Enemy, whether in Advancing or in Retreating, if the Flank Supports are aware that there are no patroles or parties detached to keep a look-out, to prevent or to give notice of the Enemy's attempting to turn the Flanks, these Supports will detach a Double File for such purpose.

28. Should the Enemy out-flank a line of Skirmishers, suppose the Left, the Officer commanding the Left Support will advance obliquing to the Left; and, when clear, will front, turn, extend from the right, dash up, and prolong the Line of Skirmishers.

29. If a Flank is turned, suppose the Left, the Officer will wheel up the Left Support towards that Flank, Advance, and resting his Right on the Left Flank of the Line of Skirmishers, will extend either perpendicular to the old line, or will throw forward his outward flank in such oblique direction as the case may require.

30. Should it be deemed necessary to meet a Flank Attack by forming a new Line to that Flank, it may be expeditiously done as follows:—the Wing of the Skirmishers on the Flank threatened will be wheeled up into the direction required, either oblique or direct, as the case may be. Its Flank Support will also wheel and dash up, extending on the march to prolong the new Line, while at the same time the other wing of the Skirmishers will close to its inward Flank, and move with the remaining former Supports to occupy their respective positions as Supports in Rear of the new Line.

NEW LINE TO A FLANK.

31. When the whole or any part of a Chain is to be reinforced, the Supports destined for this purpose will be denoted, and will proceed as the case may require, as directed, Section XII.

REINFORCING SKIRMISHERS.

32. In all cases, when Supports are ordered to reinforce Skirmishers, or sent to meet Flank Attacks, &c., they will immediately be replaced by parties of equal strength sent from the Battalion or Reserve.

33. When the Enemy's Skirmishers at particular points can-

ATTACKING IN CLOSE ORDER. not be made to give way by firing, or when from being posted on more favourable ground their fire has the superiority, also in attacking bridges, defiles, entrances of villages, &c., where it may be requisite to make a bold Attack with the bayonet, and where it is not deemed advisable that the Chain of Skirmishers should charge, the Supports formed in close Column of Sections or Sub-divisions will be brought up to the Line of Skirmishers, formed either in one or more bodies as the case may require. The Skirmishers will keep up their fire until the Supports dash past them, when they will assemble as Supports, and follow if so directed. On some occasions it may be requisite not to close the whole or only a part of the Skirmishers, in which case they will continue extended on one or both Flanks of the Column of Supports, as the case may require, and, by a lively fire, sustain the Attack.

SECTION XI.

Duty of Reserve.

1. WHEN a body of Light Troops is detached to act entirely by itself in Skirmishing Order, it must form a Reserve from its own strength; that is to say, such part of the force which is not absolutely required in the posture of affairs to form the chain of Skirmishers, and the Supports must be held together in Close Order as the "Reserve," which, being a compact body, is ready to act on any emergency, easily moved to any point most threatened, and to give effectual support to the bodies in front. From it are likewise sent out all parties either to repel or to make Flank Attacks, &c. It is obvious that this force should be concentrated in one place, because while the Supports are destined to maintain the Line of Skirmishers at various particular points, the object of the Reserve is to ensure a favourable issue to the whole engaged, so as to obviate the evil consequences which might arise from any disorder, which in prolonged Skirmishes may at times be occasioned, even from slight causes. Its most essential duty is however to act against the Enemy, when nothing more can be performed by the Skirmishers, whether from being driven in, or that it is wished to follow up some decided advantage which they may have obtained over him. With regard to the Position of the Reserve, whether it should be posted in Rear of the Centre or behind any other part of the Line, must be determined by the nature of the ground and the proceedings of the Action. In general it should be posted in the best position which will admit of affording ready support in case of any disaster, so as to be enabled to restore and maintain the Action, or, in the event of success, to follow

<small>UTILITY.</small>

<small>POSITION.</small>

Duty of Reserve. 61

up the advantage gained. Hence it is indispensable, under every circumstance, that it should have free communication with the Supports and Skirmishers. When the ground offers these required advantages, a Central Position is always to be preferred. The best point of view to observe the progress of the Attack or Defence, and the best position for Defence, in case of Defeat, are always to be considered; as also that it ought to be secured from the Enemy's fire as far as circumstances will permit, either by its distance or by being placed under cover. And could it be concealed from the view of the Enemy, it might be of some importance, as it would be enabled to act at the decisive moment of Attack or Defence more effectually as being unexpected. It is, therefore, seldom possible in Action to place the Reserve advantageously, and at the same time preserve its relative Central Position.

2. A Reserve must be constantly held in hand ready to move in any required direction, and always be prepared to throw itself into Square in case of any sudden Attack of Cavalry. It ought, therefore, for these purposes to be generally formed in Columns of quarter distance, and only in Line when earthen walls, ditches, &c., afford good cover, or if exposed to the fire of artillery. And when it is necessary to deploy, for such purpose of Attack or Defence as may require it.

3. A Reserve will, in general, conform to the movements made by the Skirmishers and Supports, and will, CONFORM TO as far as circumstances permit, preserve its relative position. Should the Skirmishers change front to a flank, or form a new line in a direct or oblique direction, the Column of Reserve will proceed to occupy such position in rear of the new Line as seems most eligible. But as a Reserve is destined to uphold the whole, and also particular points of the chain, its attention ought to be directed to, and its position determined by, the general posture of affairs; therefore its movements cannot always exactly depend like those of the Supports on those of the Skirmishers. It will, however, change from one position to another, when the original relative position to the chain of Skirmishers is by the movements of the latter so much changed that it can no longer attain the object for which it is destined.

4. In offensive operations, when our Skirmishers have driven in those of the enemy, the Reserve will proceed to attack his main body, and carry his position.

5. If our Skirmishers are overpowered and driven back by a superior force of Infantry, the Reserve will deploy to protect them, and repulse the attack. If the Enemy's Cavalry threaten an attack, the Reserve in column will move up ready to form Square, or to sustain and bring off the smaller Squares of Skirmishers and Supports, as directed, Section XIII.

6. The general operations required of Reserves of a body in Skirmishing Order, will be found under various heads afterwards given.

7. A Reserve on some occasions may consist of a different description of Troops from what the Skirmishers and Supports are composed of, such as a Battalion or Brigade of the Line, and under particular circumstances, sometimes it may be formed of Cavalry. Although one main Reserve is in general sufficient, it may happen that from the great extent of a Line of Skirmishers, engaged in woods or broken ground, and from circumstances depending on measures necessary to be taken from the position of affairs, that more than one Reserve may be required. And in cases when villages, woods, &c. are to be attacked, when, exclusive of the main body, Companies are ordered to make separate attacks at particular points, they will leave part of their Supports in Reserve, or such local Reserves may be formed of other Companies. These in Continental armies are called "Replis."

COMPOSITION OF.

LOCAL RESERVES.

8. All flank parties or patrols sent for the purpose either of protecting the flanks of a line of Skirmishers, or of a Column when on the march, &c., are always detached from the Reserve; and when acting on the defensive or in position, additional parties may be sent to harass the Enemy's flanks, which may prove a successful means of retarding his advance. Likewise when acting on the offensive, all parties sent to act on the Enemy's flank, either to out-flank or to turn them, are detached from the Reserve When Supports are ordered to reinforce the Skirmishers, to prolong the Line, or employed for any other purpose, equal parties will always be sent from the Reserve to replace them.

A Reserve ought likewise to be on the alert that its own flanks are not turned, or itself attacked. Indeed there has been recently two instances of the kind, which occurred in the late affairs in Affghanistan.

SECTION XII.

INTERNAL CHANGES AND COMBINED MOVEMENTS.

Relieving Skirmishers.

1. WHEN Skirmishers have suffered severely by casualties from being long under fire in a protracted action, or are much exhausted by moving over broken ground, it may be necessary to relieve them; although it is not very advisable to do so when they are actually engaged, unless they have spent all their ammunition. The manner of executing this change depends on the situation they may be in at the moment the relief is to take place, *viz.*, whether they are Halted,—Advancing, or Retreating.

2. When a Line of Skirmishers is to be relieved while halted, the Officer commanding will take an opportunity of doing so when it is under cover, such as behind a fence or in strong ground. On the sound " Relieve Skirmishers," the Supports extend in the rear; while advancing, taking care to cover the same space of ground as that occupied by the Line to be relieved. The whole then dash up, each File moving upon one of the old Line which it is to relieve, and takes post on its right, and open their fire. Each File of the former instantly runs to the rear, and when out of reach of the enemy's fire, they will close and form up in support in as many parties as the others consisted of, each portion closing on the march to their proper right, left, or centre, as the case may require.

HALTED.

3. But should an immediate advance be intended, the old Skirmishers ought to remain in the Line, lying down, instead of exposing themselves to a fire whilst retiring; and when the new Line has advanced a sufficient distance, they will close and form the Supports.

4. If the relief takes place while advancing, the Supports extend, and run up in the same way; but in this case pass briskly in front of the old line, and instantly proceed skirmishing. The old Skirmishers lie down, and will not rise up until the new chain has advanced sufficiently to protect them. They will then close, and form Supports as before.

ADVANCING.

5. In relieving Skirmishers, when either halted or advancing, should there be enclosures, hollow ways, &c. to be passed, as also in defending villages or such places where a sufficient passage cannot be obtained for the extended files, the force or Supports destined to relieve the Skirmishers may move up in close order under cover, so as not to be exposed to fire, and then extend to relieve the old Line. When this is accomplished, the latter will close in, and withdraw in the same manner.

6. When Skirmishers are to be relieved while retiring, the Officer in command will select a good position for the Supports at a considerable distance in the rear, where they will extend, each file getting under cover, and lie down. The old Skirmishers continue to retire in their usual order, until within 20 or 30 paces of the new line. They then run through it to the rear until out of range of fire, and then close in to form Supports as before directed. The instant their front is cleared, the new Skirmishers will open their fire, but ought not to commence retiring until the old Skirmishers have re-formed as Supports, particularly as being a fresh body they may, if it should be required, be able to keep the enemy in check for a short time. They will afterwards continue the retreat in the usual manner.

RETREATING.

7. If during a retreat both Supports and Skirmishers are to be relieved by fresh parties from the Reserve, or other body of Troops, such parties will be previously posted in their proper order in position. The old Supports and Skirmishers will retire through them, and then form on the Reserve or Column, as the case may be.

8. Either of these two last ways is the best mode of relieving Skirmishers, when acting as a rear-guard, as they can, on arriving at good defensive positions, defiles, &c., take post, and relieve one another in succession.

9. But on common occasions, if the Supports and Skirmishers are all relieved by fresh parties, each of the Supports preserves the relative position, with respect to their own Skirmishers until the two lines have relieved one another.

A Line of Skirmishers reinforced.

10. A chain of Skirmishers or any part of it may be reinforced, either by strengthening it or by prolonging the Line to one or both flanks.

11. When a Line is to be strengthened, the Officer commanding will, as he may deem necessary, order the whole or such portion of the Supports to reinforce the whole or any particular part of the Line he wishes. On the sound "Reinforce" the Supports, or such portions as may be ordered, will rush up, extending on the march to the post to be strengthened, where they will mix with the others; each File getting between two of the old Line, dividing the distance.

CHAIN STRENGTHENED.

12. But if it is wished to avoid mixing the Files, the old Skirmishers, if not too much exposed to the Enemy's fire, may be ordered to close their distances to a Flank or the Centre, as the case may be, and the reinforcement will fill up the vacated ground.

13. Should the Enemy out-flank our Line of Skirmishers, or should it be wished to out-flank that of the Enemy, our Line may be prolonged to either one or both Flanks, as may be necessary. Suppose to the right, on the sound "G. and reinforce," the right Support will advance, inclining to the right, and when clear of the Flank of the Line six paces, will front, turn, and extend from the left, dash up to prolong the Line.

CHAIN PROLONGED.

14. In these cases the Supports will be replaced by parties from the reserve.

15. A Line may be weakened by withdrawing a portion of the Skirmishers from the Line; or it may be diminished by a portion being called in from one or both Flanks.

WEAKENED.

16. When a portion is withdrawn from the general Line, the Files ordered to retire will run to the rear, and close on their respective Supports. And if necessary, the remaining Skirmishers will extend distances from a named File so as to cover the vacancies, and occupy the breaks in the Line. This is particularly applicable to again withdrawing Skirmishers who had reinforced a Line.

17. When a Line is to be diminished and a part recalled from a Flank, the Section, Sub-division, or Company so denoted by signal, will face about and retire, closing on the march, and join the Support in rear of

DIMINISHED.

Changing Front.

18. A line of Skirmishers may change front to the right or left, on a flank or centre file.

19. If to the right, on the sound "Right Wheel," the right file faces to the right, kneeling. The others rise up, half-face to the right; each file moves in double time direct to its place in the new line, preserving distance, and taking up dressing from the halted flank, form in succession on the right file, and kneel down.

20. If the right is to be thrown back, the left file faces to the right, kneeling. The others rise up, face three-fourths to left-about, each file moving in double time to its place in the new line, taking up dressing and distance from the next halted file, halt, front, and kneel down.

21. When the change is to take place on a central file, and one wing is to be wheeled up and the other backwards, if to the right, the left wing will wheel up, and the right one back, as above directed.

22. These are necessary movements for parade practice. But on service it can rarely be required that an extended line when actually engaged should change front direct to a flank, because time would be lost, the quantity of fire diminished, and the outward men much exhausted by making so extensive a wheel.

23. It may however be frequently necessary when in action to throw a wing forward or backward in a greater or less degree.

24. On common occasions, if a wing is to be thrown forward, suppose the right, on the sound "Right Shoulders forward," the left file half faces to the left. All the others step off, preserving distance from the inward flank; each gradually increases the pace in succession towards the outward flank, which moves in double time, and to which they must look for dressing, and bring forward the shoulders conformable to its progress.

SHOULDERS FORWARD.

25. But when actively engaged, the fire may be continued during the movement, by the flank file being faced to the required oblique direction, and the next file or two dress up into the new line. All the others will conform in succession, preserving distance from the inward flank, and each opening their fire as they arrive in their places.

Skirmishers changing Front.

WING THROWN BACK. 26. If a wing is to be thrown backward, the flank file will be faced into the required direction; all the others go to the right about, and proceed as directed for a wing thrown forward until they gain the new direction, when they will halt, front, &c. But if engaged at the time, each file as it reaches the new line will halt, front, and open its fire.

IF BY ALTERNATE RANKS. 27. If a wing is thrown back whilst firing and retiring by alternate ranks, the rank whose turn it is to retire may, after passing the rear line, bring shoulders forward in such direction as may be required, and then halt, front, and load. The next rank, after firing, retires in the same manner, and they thus continue until fresh orders are given.

28. During all these movements of throwing a wing backward or forward, should the "Halt" be sounded, the whole will halt, each file correcting dressing and distance, and gets under cover.

29. If the "Advance" or "Forward" be sounded, the whole will take up the "Quick Step," and move off together direct to the new front.

Skirmishers closing to a Flank and changing Front.

30. A line of Skirmishers may close to a flank or the centre, forming Sub-division or Company, as the case may be, to right or left, either forward or backward. If to the left forward, the whole face to the left, the left file stands fast; the others close in double time, inclining a little to the right, and form line in succession on the right of the halted file; each rear-rank man marking time two paces, to allow the front-rank man to get into his place.

31. If to the right forward, the whole face to the right, close, and form on the left of the halted or right file. But if to the right backward, the right file faces to the left; all the others face to the right and close. Each file in succession halts, fronts, and forms on the left of the standing file, each front-rank man marking time two paces to allow his rear-rank man to get into his place.

32. When a line is to close on the centre and form line to the right or left; if to the left, the centre file will face to the left. All the others will face inwards. The right wing will proceed as directed for closing and forming line to the left

forward; and the other wing will act as directed for forming to the right backward.

33. These movements combine the advantage of a line of Skirmishers being enabled to close and to wheel or change front to a flank at the same time; and are particularly useful when a new line of Skirmishers is formed to a flank, because by this means the old Skirmishers are readily formed into Supports.

34. When a chain of Skirmishers is covering the movements of a Division or Brigade, should a change of front be required, See Her Majesty's Regulations, Part V., Section IV., No. 21.

35. Line to a Flank either direct or oblique. On service when it is necessary promptly to meet a flank attack threatened by the Enemy, or rapidly to follow up his hasty retreat towards one while engaged, and that it is requisite to form line to such a flank, the whole chain may bring forward their outward shoulders, or a new line may be thrown out from the Supports or Reserve. But as either of these modes would occasion loss of time, and the latter stop all firing during its execution, the change necessary may be more speedily effected by throwing forward the wing of Skirmishers on the flank threatened, or to which the change of front is to be made, in such direction as may be required, either direct or oblique, as the case may be, and the fire continued. While its Support will also wheel up, advance, and extend to prolong the line, and open its fire. The other wing of the Skirmishers will at the same time close to its inward flank, forming line to the same, and will, with its Support, move forward to form the Supports to the new line.

36. Suppose the four front Companies are thrown out to cover a Column (Nos. 1 and 2 acting as Skirmishers, and 3 and 4 in Support), if a change of front is required to the left flank, on the sound "G. G. G." left form line, No. 2, or left wing of the Skirmishers, wheels up to the left. Its Support, No. 4, also wheels, and moves forward extending to prolong the line to that flank; while No. 1, or right wing, closes, forming to its left, proceeds to form the right support, and No. 3 having wheeled, moves to form the left one.

37. Should an oblique line be only required, and if to the left, the left flank file of the left wing will be faced OBLIQUE into the required direction, and the next file or two LINE. dressed up; thus a base will be given for the others to form upon. The whole will then proceed as above directed, conforming to the new direction, and making due allowance

for the obliquity of the new line. On these occasions the Officer commanding the different divisions, whether Skirmishers or Supports, must take care that these bodies wheel with celerity all at the same time, and move rapidly into their new positions. Should the chain consist of three or more Companies, the new line to a flank may be formed on the same principle by wheeling up two Companies of it into the new direction, &c. &c.

38. On service, when flank parties are out, they will likewise conform and protect the flanks during the movement.

SECTION XIII.

Rallying.

1. RALLYING is when Troops in skirmishing order may have by any accident got into confusion, are again restored to fighting order; or when an extended line is to assemble and re-form in close order. On such occasions, in the field, the speediest mode of forming and presenting a front should be the first object. The inversion of files or ranks is not to be attended to at the moment if time is gained. Hence if by any untoward accident Skirmishers have been thrown into confusion, or when, by a protracted combat over a diversified country or broken ground, the Skirmishers of different Companies or other body have got mixed, the senior Officer will get three or four files formed as quickly as possible, placing them in the direction required, so as to give a base to the new line and as a point of formation for the others to rally upon. The nearest Officers and Non-commissioned Officers will form their men on these files, which will be taken up in succession to the right or left, or to both flanks as the case may require. The men may afterwards, when circumstances permit, get into their proper places and be told off again. Nothing distinguishes well-trained Light Troops more than when the Officers have their men always well in hand, and that the Skirmishers give a prompt attention and ready obedience to orders or bugle sounds while under fire.

RECOVERING ORDER.

2. No rallying or assembly ought to take place within the effective reach of the Enemy's fire unless in cases of emergency. When Skirmishers are to be withdrawn, they ought first to run back out of reach of fire, and then close and re-form, as in the manner prescribed for Skirmishers when relieved, either halting or retiring.

3. It is of the utmost importance that Skirmishers covering Troops about to engage, or the manœuvres of a Battalion or other body, as soon as the formations are completed, should clear the front with the greatest celerity.

4. Troops may be re-assembled from extended order in various ways according as circumstances may require. On the usual occasions of parade exercise, if Skirmishers are to assemble on the line they hold, they will proceed as directed.

If by the " Close," they will rally on their Supports, &c.

If by the " Assembly," both Skirmishers and Supports retire, &c.

If only a portion of the Skirmishers is to be recalled, the part denoted will proceed as directed. Section XII. No. 17.

5. In the field, when Skirmishers are covering Lines or Columns, and when the decisive moment arrives for these either to make an attack or to defend themselves, the Skirmishers will be recalled either by " Order," or by the " Assembly." On this signal the Skirmishers and Supports will instantly face about, and, without waiting for one another, will rapidly retire, while the Officers will endeavour to discover the situation of the Troops, and adopt that mode of clearing the front which will least impede their movements and soonest leave them free to fire or to advance. This is generally done by the whole making the best of their way to the rear, through the intervals between regiments in Brigades or Columns; and only the most outward sections of the Skirmishers going round the flanks. The showy mode often practised at field-days of withdrawing all towards the flank of lines, would, on service, be neither proper nor practicable. It frequently happens when the Enemy makes a decided attack on the Troops posted in position that the Skirmishers will be driven in close to the line, in which case they must get through it to the rear as they best can, some files opening at intervals to let them pass. But should they be so pressed as not to have time even to do this, in such extremity the Skirmishers ought to throw themselves flat on the ground, to unmask the line, and permit its firing over them. This may likewise be done on some occasions when Troops are advancing; and when the Troops have passed, they will close, and proceed to such place as may be appointed.

SKIRMISHERS RECALLED.

6. Should Skirmishers be suddenly overpowered by numbers before their Supports have time to reinforce them, they

will retire and rally on their supports, form in line with them, and will conjointly endeavour to check the Enemy, or make good their retreat, if timely aid is not at hand to succour them.

7. When Skirmishers on being recalled have passed through Lines or Columns to the rear, they will there close and form up, ready to proceed to such point as may be directed.

Resisting Cavalry.

8. Although Light Troops in a close country are greatly an overmatch for Cavalry, and if properly employed not often exposed to be charged by them, because Cavalry in such ground seldom rashly commit themselves, or venture to attack Skirmishers who show a disposition to make a determined resistance, yet Light Troops must at all times be prepared to throw themselves into Square from every situation, whether acting as Skirmishers, Supports, or Reserve.

9. In skirmishing across a country, either in advancing or in retreating, particularly while passing over any open space or places having heights, hollows, coppices, &c., where Cavalry might be concealed, an attentive look-out must be kept to guard against a sudden rush of Cavalry upon the extended line. In such situations, and indeed at all times, Skirmishers ought to be jealous of Cavalry on their most distant appearance, and to be constantly on the "alert" to prepare to receive them before they get too close, so as never to allow themselves to be individually charged by them. Therefore on the first appearance of Cavalry the alarm will be sounded to put the men on their guard, and the Skirmishers to keep a sharp look-out.

PRECAUTIONS.

10. Should the Cavalry press on, the supports will advance towards the threatened points, looking out for favourable situations to form Squares, choosing such positions that they may fire clear of one another, while the reserve will at the same time follow up and form square, ready to sustain the smaller ones, and interpose itself if necessary between them and the Cavalry.

11. On the sound "Form Square," the Skirmishers will instantly retire on their Supports advancing to meet them, and form Square with them by Subdivisions or sections, as directed by Regulations, Part II., Sect. XXIV. They will take care while running to the rear to open to

SQUARES.

right and left, so as to leave the front of the Supports clear to open their fire when necessary.

12. Should there be good cover near the line of defence, such as a hedge, ditch, copse, garden walls, &c., the Skirmishers, or such portions as are nearest such objects, particularly the distant files in the outward flank of the line, will at once make for such places, and aid the squares of Supports by a cross fire to gall the enemy and check their advance, (Regulations, Part V. Section 4, No. 34,) while the Officer will take care that the men are so placed as to fire clear of one another.

13. The Supports are only to be considered as points of formation for the Skirmishers to form Squares upon, when there is time for the chain to run back upon them and form without risk or confusion, so as never to interfere with the Supports opening their fire. Reg. 281, No. 31.

14. Therefore, when the chain is so suddenly threatened that there is not sufficient time to reach their Supports, the " Alarm " and " Assembly " will be sounded. (Reg. Part V., Sec. 4, No. 33.) The Skirmishers will rapidly close, each Wing or Company, as the case may be, to its own Centre, or other point denoted, and form Rallying Square or Circle round their respective Officer, who will hold up his sword, and who will take care to choose such positions for their Formation as will be in Echellon of the Squares of Support, so as not to cloud their front, while a few Files of the Skirmishers should, if opportunities offer, take advantage of cover to take the Enemy in Flank. These small masses will profit by every moment of respite the Cavalry allow them to draw close to one another so as to unite, and then to put themselves in security by retiring, and endeavouring to gain the Squares of the Supports and Reserve; or at all events to gain some favourable position for Defence. For this end the Officers will form them in close column of Sections, and move off, and if again threatened, they will halt and re-form Square.

RALLYING SQUARE.

15. On all occasions small Squares, whether formed of the Skirmishers alone, or of these with their Supports, must obstinately defend themselves until the Reserve, or their own Cavalry, come up to assist or relieve them. Men formed in this manner may with firmness resist a considerable body of Cavalry, of which the last wars offer several memorable instances.

16. Should the attack be so unforeseen that the whole or part

SURPRISED BY CAVALRY. of the Line of Skirmishers have not time either to take advantage of cover, or to form Rallying Square, they must, in the first instance, form small groups by three or four Files getting together, forming round an Officer or Non-commissioned Officer, if at hand, standing back to back, and with the greatest coolness pay attention to the movements of the Cavalry. The Front Rank will make a half face to the right, and hold their Muskets nearly as if coming down to the Charge; only the point of the Bayonet raised opposite the left eye, and thrown forward ready to parry a thrust or cut, or to make a lunge at the horse. The whole will reserve their fire for the last extremity; else, by throwing away their fire, the Cavalry might dash in with Lance or Sabre. They will take every opportunity of uniting and forming into larger parties; and such parts of the chain as are not immediately attacked will proceed to succour and unite with those which are nearest them, or in most danger. Even if Cavalry dash on, small squares or parties should avoid firing, but firmly await the attack, until the Cavalry either disperse or turn to retreat; a portion may then fire with effect, but a part should hold their fire in reserve, lest the Enemy might return to the charge. However, if it becomes necessary to fire, it ought not to take place until the Cavalry are so close that they may be certain of hitting them and frightening the horses.

17. After the Cavalry have been repulsed and driven off, if no fresh dispositions are ordered, a chain of Skirmishers will be re-established on the former line of defence.

SQUARES REDUCED.

18. Detached Squares composed of Skirmishers and Supports may be reduced either by the "Skirmish," "Close," or "Assembly."

19. If by the "Skirmish," the former Supports, as being a fresh body, will dash out, extending on the march as Skirmishers, and cover the ground occupied by the former ones, while those that were the old Skirmishers will remain and become the Supports.

20. If by the "Close," the Squares will form line, and retire in double time to the nearest Flanks of the Troops in the rear.

21. If by the "Assembly," the Squares will break, and run in in dispersed order, through the intervals, or by the nearest flanks to the Rear, and then form up.

Rallying.

22. Should the Enemy's Skirmishers or parties of Cavalry hover round a Square or Column in quarter distance, whether halted or in movement, Skirmishers may be thrown out to keep them at a distance, and prevent their harassing it. Should the Cavalry charge when the Column forms Square, part of the Skirmishers will lie down all round close under the bayonets of the kneeling ranks, ready to dash out when the Enemy turns to move off again. The remainder may form in the rear, and watch every opportunity that may offer of quickly filing out, and fire on the Enemy from whichever flank seems best; or if they could have gained any cover near at hand, it would be preferable.

COVERING A SQUARE OR COLUMN.

23. As an example how small bodies of Infantry, acting with coolness and courage, may resist considerable bodies of Cavalry, the following one of a Detachment of French, (consisting of 70 Grenadiers of the 22nd Regiment, by the French statement.) under Captain Guache, may be given, as recorded by Colonel Napier in his History of the Peninsular War. During the operations of General Crawford on the Coa (when Massena, with a superior force, covered Cuidad Rodrigo) he took a position in front of Almeida. The French pushed on foraging parties to the villages of Barquillo and Villa del Puerco, in his front. He endeavoured to cut them off, and placed two ambuscades for this purpose. About day-break, on the 11th June, 1810, two French parties were observed, the one, Infantry, near the Villa del Puerco, the other, Cavalry, at Barquillo. The open country on the right would have enabled his Cavalry to get between the French Infantry and their point of retreat; but as this was circuitous, Crawford preferred pushing straight through a stone enclosure as the shortest road. This fence proved difficult, the Squadrons were separated, and the French, 200 strong, had time to draw up in Square on a steep rise of ground, but so far behind the ridge as not to be seen until the ascent was gained. Two Squadrons of the 14th Light Dragoons, which led, galloped in upon the French Square, commanded by Captain Guache, which stood firm, received the charge, then opened their fire in front and both flanks, and thus foiled the Dragoons. Colonel Talbot soon afterwards followed with four Squadrons, and immediately bore gallantly in upon Captain Guache; but the latter again opened such a fire that Colonel

Talbot and 14 men fell close to the bayonets of the Enemy, who resumed their retreat, and effected their escape."*

24. They, however, ought to avoid taking shelter on the banks of impassable rivers, or under precipices, under dead walls, &c., as the Dragoons might hem them in under such obstacles. And instances are not wanting, in such circumstances, when a portion of the Dragoons have dismounted, and by ascending rocks or precipices, have fired down upon the Infantry so surrounded, and thus compelled them to surrender.

25. Napier also gives another instance. When General Hill made a demonstration from Portugal into Estremadura by Albuquerque, 28th December, 1811, in order to create a diversion in favour of the Spanish armies, his Advance of Cavalry fell in with 300 French Infantry, under a Captain Neveux, and a few Hussars at Nava de Membrillos, part of a foraging party. This small body retreated on Merida, closely followed by 400 of the Allied Cavalry, who had orders to make every effort to stop their march. But, to use the words of General Hill, the intrepid and admirable manner in which the Enemy's Infantry retreated, formed in Square, and, favoured by the nature of the country, of which he knew how to take the fullest advantage, prevented the Cavalry from effecting anything against him. The able Officer who commanded reached Merida with the loss of only 40 men, all of whom were killed or wounded by the fire of Artillery.†

Squares.

26. On the appearance of Cavalry, the "Alarm" will be sounded, to put the Skirmishers on their guard, who will keep a sharp look out. The Officers commanding the Supports will immediately form each into Close Column of Subdivisions, looking out for good positions, and be ready to move towards the Skirmishers.

27. But on a sudden rush of Cavalry, when Skirmishers have not time to fall back upon their Supports, RALLYING each Wing or Division of the Skirmishers, as the SQUARE. case may be, will form *Rallying Square* on its own centre, as directed in Her Majesty's Reg., Part II., Sect. 23, No. 1, p. 63; and Part V., Sect. 4, No. 31, p. 281.

---

* See Napier, vol. iii., p. 284.
† Ibid., vol. iv., book xv., p. 326.

28. And in this case each Support will form Square, as directed for a Company; taking care to choose such a position as will enable it to fire clear of the Rallying Squares, viz., the Column of Sections will close to the front. The two rear Sections face about, the two outward Files of the centre Sections face to the right and left outwards, as directed in Her Majesty's Reg., Part II., Sec. 23, No. 4, p. 65; and Part V., Sect. 4, No. 32, p. 281.

SQUARES OF SUPPORTS.

29. But if there is time, and that the Skirmishers fall back on their Supports, on the "Alarm" sounding, the Support (in Close Column of Subdivisions) nearest the threatened point of attack will advance first; then the next, and so on towards the Chain, in order that they may succour the Skirmishers, and also be sufficiently in Echellon to fire clear of one another when the Squares are formed.

SQUARES OF SKIRMISHERS AND SUPPORTS.

30. On the "Assembly" or sound "Form Square," the Supports will Halt and wait for the Skirmishers, when they will form Square together as directed for two Companies, by column of Subdivision closing to the Front and proceeding as directed above for Sections. In this case, as soon as the Skirmishers have joined and formed close up in the Rear, without regard to the individual places of the men, the whole will prepare to receive Cavalry.

31. If a Reserve, consisting of three or four Companies, has to form Square, it will be done as directed for the wing of a Battalion, viz., Columns of Subdivisions will close to the Front. The two Front ones stand fast; the two Rear ones go to the Right-about; the Centre ones wheel outwards by threes, and each close up to their respective Fronts to form the Side-faces, as directed in Her Majesty's Reg., Part. III., Sec. 12, No. 3, p. 118.

SQUARE OF RESERVE.

32. When a Reserve, consisting of five, six, seven, or more Companies, has to form Square, it may be done by quarter distance column, as directed in Her Majesty's Reg. Part III., Sec. 21, No. 1, p. 144. Or the Square may be formed if from Line on the two Centre Subdivisions; or on the two Centre Sections, when it consists of three or four Companies, as directed in Her Majesty's Reg., Part. III., Sec. 21, No. 5, p. 147.

BATTALION SQUARES.

33. Squares may be formed by a Battalion in Open Column, on a Front or Central Company, whether Halted or Filing to a Flank, as directed in Her Majesty's Reg. Part III., Sec. 29, No.

1, p. 161. A Battalion moving in Echellon will form Square as directed in Her Majesty's Reg., Part III., Sec. 45, p. 191.

34. The French and other foreign troops, owing to their being, on all occasions, formed in three deep order, have a great facility in forming small detached Squares. A Company acting as a Support has only on any sudden emergencies to wheel back the outward Flanks, and thus form a Circle, having always one Rank ready to kneel down, and two to fire, with room in the Centre for Supernumeraries.

35. If there are two Companies in Line, a named Subdivision stands fast, the others have only to face about, and bring right and left shoulders forward to complete the Square.

If in column, the front Company wheels back the outward Flanks a few paces, and the rear one wheels up in like manner.

36. If there are three or more Companies in column, the front one stands fast; the others, except the last, take subdivision distance, and wheel up by subdivisions outwards to form the flanks. The rear one closes up and faces to the right about.

37. These are all simple in their formations, and easily reduced when required.

38. A single Company, or two, when in sections of threes, have only to close on the leading one, and halt front in that order, and then proceed as above.

39. If three or four Companies are in column, they need only form Sections of threes to a flank, close up and front in that order. And the centre one or more taking one-third distance, may wheel up by subdivisions outwards; while the rear one after closing up, face to the right about.

40. As an example how a detached body of Infantry may act and successfully resist Cavalry, the following may be given.

In the affair of El Boden, 24th September, 1811, Sir Charles Colville's Brigade, consisting of the 5th and 77th Regiments with the 21st Portuguese and two Brigades of Artillery, being in Advance, were attacked by an overwhelming force of French Cavalry and Artillery, under Montbrun. The Enemy having taken the Portuguese guns, Major Ridge, with the 5th Regiment, most gallantly made a sudden dash at the *Enemy's Cavalry* and actually recovered the lost guns. The destined Supports not coming up in time, these Regiments were obliged to Retreat, which they did in the most admirable manner, formed in Squares. And although surrounded and pressed on all sides by multitudinous Squadrons, they continued alternately retiring and repulsing the Enemy with the most determined resolution and with a firm and even step, and thus proved how fruitless to match the sword with the musket, to send charging Horsemen against the steadfast veteran.*

The invincible resistance made by our Infantry at Waterloo against the most renowned Cavalry of France is too well known to be repeated here.

Skirmishers resisting Horsemen.

41. When Skirmishers are actually surprised by Cavalry, and that they cannot avail themselves of any of the means of defence above specified, each one in such emergency must shift for themselves by endeavouring to get up banks, behind trees, rocks, or to gain the nearest fences or other cover, and from behind these keep up a well-directed fire; at the same time taking the utmost care to fire clear of one another. But should even this be impracticable, from want of time or cover, and escape impossible, they must preserve great self-confidence and presence of mind, and each individually defend himself with coolness and courage. They will find that these qualities, with judgment in applying the instruction hereafter detailed, will give them an advantage in this seemingly disadvantageous struggle, and that they can combat such Enemies with success. It ought to be impressed on the men, that when they have to defend themselves against Cavalry, experience has proved that a single Horseman, even on open ground, has no superiority over an Infantry Soldier who acts with cool determination; and that in broken ground, the Infantry have a decided advantage, because from animal instinct the horse has an innate shyness for strange objects and such encounters, which is increased by his faculty of recollecting former dangers, and the dread of the report of fire-arms. This makes him unmanageable, and prevents the rider from guiding him easily against a Skirmisher, who moves nimbly about; particularly as the shape of the horse hinders him from turning so quickly as a man. And with regard to the Dragoon himself, his strength consists more in his boldness

OVERPOWERED BY CAVALRY.

INDIVIDUAL DEFENCE.

ADVANTAGES OVER A DRAGOON.

---

* See Napier, vol. iv., book xiv., p. 242.

and individual adroitness, than in the use of his arms; for neither the sabre nor the lance are very formidable to a brave Infantry Soldier, experienced in the use of his bayonet; and the effect of fire arms, whether a pistol or carbine, are too uncertain from a person on horseback, to be taken much into consideration or to be apprehended.

42. Every soldier must, in the first instance, take care that his arms are loaded, and that he never discharges them until every means of defence afforded by the use of his bayonet are exhausted. He must reserve his fire for the utmost necessity, and the moment it will be effectually certain of hitting. He ought never to allow himself to be induced to fire by any attempt of an Enemy to make him do so, because an experienced Dragoon will not at once approach an Infantry Soldier who has his musket loaded. He will first keep moving about at some distance, firing his carbine or his pistol in order to induce the Infantry Soldier to discharge his. In such case he may deceive the Dragoon by presenting at him the same moment the other fires, and then instantly go through the motions of loading, to make the other believe that he has thrown away his fire.

RESERVE HIS FIRE.

STRATAGEM.

43. He must be made sensible of the advantage he has if he gains the left side of a Horseman, which is his weakest point. Should a Dragoon, either with pistol or sabre in hand, ride directly at him, he will allow him to approach within a few paces, and then spring a step or two to his own right to avoid the horse, and thus gain also his left side; so that the rider can either do nothing, or can only reach him with uncertainty, during which he may find an opportunity of shooting the rider; or of running his bayonet into the horse's side which may bring him with his rider down, or at all events make him restive and run away. Indeed every endeavour must always be made to maim the horse with his bayonet rather than the rider; and he ought never to fire at the horse, however great the probability of hitting him, because, unless the animal was mortally wounded, he would still be able to move about, which might prove of advantage to the rider, because he himself having fired, would be unloaded and consequently without his last resource of defence. He must avoid holding his bayonet towards the breast of a horse advancing on him, because, from the formation of its body it cannot perceive the

TO GAIN THE LEFT SIDE.

danger, but would rather run upon it, and thus throw him down.

44. If a Skirmisher advances himself against a Horseman, he will give the horse a rap on the left side of the nose, which will cause him to swerve to the right, and so that he will gain his left side, and be enabled to stab the horse or fire at the rider. In such case he will apply, as circumstances and opportunities may require, either of these means which give him the superiority over the horse or the rider. Should he have missed fire, he must prepare for single combat by parrying off with his bayonet, any blow aimed at him by the dragoon, and always endeavour to strike the horse on the nose, to gain his left side, and in this case thrust at horse or man as he best can. Should he by this means get momentarily rid of them, he will load, and prepare to receive the Dragoon as before.

45. He must recollect that a Dragoon who acts with prudence, will charge him vigorously, and that when within ten paces he will oblique to his left and then to his right in order to maintain his own right against the Infantry Soldier, and to keep him on that side; and that to insure this, he will probably keep making a circle round him. This is the most critical position for the Infantry Soldier, because the movements of the horse may confuse him. In such cases he must endeavour to keep two or three paces from the horse, moving about and parrying, but always watching an opportunity of making a thrust. If he feel tired, he ought to stand still, and turn about on his left heel as a pivot, keeping his arms at the guard as nearly as if coming down to the Charge. If he wishes to get out of the circle in which the Dragoon has placed him, he will endeavour, by a rapid dash, to get directly in Rear of the horse, and thrust either at him or the rider. If he has not touched the rider, and if the Dragoon attempts again, by making a circuit, to gain the right again, he ought to try to get in front of him, and strike the horse on the head. If the Dragoon, instead of making a circuit, faces to the right-about to sabre him on that side, he ought in this case to make a dash by the Rear to the left at the moment he sees the rider going to turn his horse, and make a thrust at him.

46. Light Troops may rest assured that in such combats, the struggle, in most cases, even against the most active Horsemen, will not be long undecided, and will end to the advantage of the Infantry Soldier.

47. Against Lancers, however, a combat of this kind may

AGAINST LANCERS.
in some degree be more difficult. Yet a resolute and expert soldier, who boldly faces the danger, and acts with judgment, may unquestionably come off with success. His best chance is to stand, keeping his front always direct to the horse's head, seeming as if waiting for the Attack, but at the same time ready to spring to the Right, as soon as the Lancer gets within a few paces of him. Then after doing this to face sharply to the Left, and make a dashing thrust at the horse, before the rider has time to turn him. Should he not have succeeded in gaining the left side of the rider, he must move nimbly about, and parry as well as he can; and if he hits the lance then he ought to close with his adversary, by thrusting at either horse or man as he best can; but if Troops are broken, and retire in disorder, Lancers may do much execution.

48. If he is attacked by two Dragoons who are close together, side by side, he ought to stand and wait for them, and strike whichever horse he can reach best, and spring to the Left if he has touched the Right horse, or to the right if he had hit the Left one. If the two Dragoons have an interval between them, with a seeming design to get him betwixt them, he ought to dash at the one nearest him, and strike or thrust at the horse in such a manner as will make him swerve in such a direction, that he will get in the way of the other, and stop him while he can make a thrust at the horse, or fire at the riders. By these means he may probably disconcert both, who, even if not hurt by his fire, may, from the fright of their horses, and such unexpected resistance, be rendered harmless; and the active soldier will frequently come off victorious, even in this double Attack.

TWO DRAGOONS.

49. Should the Dragoons succeed in closing on him, he must actively turn about. In parrying them off alternately, and by directing hits, and thrusting the best manner he can at the horses' heads or riders. And instances are not wanting where by coolness, activity, and proper management Infantry soldiers have put two Dragoons *hors de combat.*

50. Should more than two riders attack him, the most probable chance is, after they have fired at him, for him to fall on the ground as if hit, pretending to struggle a little, and then lie as if he were killed. A horse will not willingly tread on a man, and a rider cannot, in such a situation, do much against him.

SEVERAL DRAGOONS.

51. On all these occasions of combats with horsemen, a single tree or other object will greatly favour the Defence, if the Infantry soldier stands close up to it, and manages so as to have his musket always on the right side of it, and keep moving round, so as to have the tree or other object always between him and the rider, but if several attack him he must stand with his back to the tree and parry their cuts.

TREES, &c.

52. Example. When the Army was advancing on Victoria a Skirmish took place, 18th June, 1813, near the village San Millan, during which a party of French Hussars made a dash at our Skirmishers. One of the Hussars attacked an Officer of the 1st Battalion, Rifle Brigade, he took refuge behind a tree, round which the Dragoon chased him several times, cutting at him with his sabre. The Officer observing a rifle lying on the ground, which belonged to a man who had been killed, watched his opportunity, and picked it up, instantly regaining the tree. He then waited for the Frenchman, coolly shot him through the body, and immediately seized the horse as a lawful prize.*

53. No stronger instance can be given of what determined Infantry can do against Cavalry than the conduct of the gallant 42nd Regiment, in the Battle of Alexandria, in Egypt, 21st March, 1800, when the French Cavalry having got in amongst them before day-light, the Highlanders fought hand to hand with the Enemy's horsemen; at times standing back to back or shoulder to shoulder, they most successfully beat them off. And in 1745 their progenitors, with only their bare broad swords, at Prestonpans, attacked not parties of Cavalry, but two Regiments, with such impetuosity that they overthrew them, and drove them in disorder from the field.

BATTLE OF ALEXANDRIA.

54. In the above instructions the worst cases that can happen have been supposed. In the field Skirmishers will seldom be left so destitute as to act singly and individually. Each will in most instances have a comrade to support him, and who will act in co-operation with him. And generally two, three, or more Files, under the most desperate circumstances, may get together, so that by the different parties being some more forward and others backward, will be enabled to fire clear of one another, and take the Cavalry in Flank by a cross fire.

* See Surtees, Twenty-five Years Rifle Brigade, p. 200.

55. It must be allowed that if Troops have been repulsed, and retire in disorder, or who may have got into confusion or be dispersed, should Cavalry, especially Lancers, get in amongst them, they may do them infinite mischief. If there are no fences or cover at hand, their only chance is to Rally sufficiently to make a bold front, and, either by offering individual resistance, or getting together in groups or masses, as circumstances may admit, make a stand to resist them and check their career.

56. Every Officer in command of a Patrole, or a party detached on any particular service, ought always to appoint a place of rendezvous, where, in case of the party being obliged to disperse, each man will make the best of his way by the safest route, and there re-assemble.

RENDEZVOUS.

SECTION XIV.

Covering Movements.

1. WHEN Light Troops are attached to a Brigade, Division, or other body, they will, when on the March, if advancing, lead the Column, or Cover the Front as an Advanced Guard. If retiring they, or a sufficient portion of them, will Cover the Rear. When the Troops form up either for Attack or Defence, whether in Lines or Columns, the Light Troops, if not required to Cover the Front, will in general be posted on the Flanks, or such position as may be required, or so disposed in Rear of Brigades that they can readily make their way through the intervals when required. By these arrangements they are (especially at Field Exercise) more in hand, all formations more rapidly made, and changes of position to either Flank easily executed; which could not otherwise be so well accomplished were they all to be kept united on one particular Flank or Point. When required to act, the whole, or only such portions as may be ordered, will proceed to the Front, or in such direction as may be necessary, and when at due distance, or when advantageous Cover may offer, will be thrown into Skirmishing Order. When posted on the extreme Flanks of a Line, or of Corps in contiguous Columns, the Light Troops, when halted, will always occupy such strong points as the ground may afford, and will throw back their outward Flank and extend an outward Section in such direction as will protect the Flank from being turned. Should the Troops change position, such portion as is not required to cover the formation, as Skirmishers, will form a Flank Patrol, or make such other disposition as will best protect it during the Movement, and, on arriving in the new position, will take post as before. They

will, likewise, when two Lines relieve one another, whether advancing or retiring, always remain with that Line which is nearest the Enemy. Thus, if posted on the Flanks of the first Line, and if the second Line passes through to the Front, they will join the second Line, and advance with it. Or if attached to a Line retiring, on coming to the Line which is to be passed through, they will halt and remain with it.

2. When no separate Light Corps is attached to a body of Troops, the Light Infantry Companies of one or more Brigades, are usually embodied to act as a Light Battalion. Should a whole Regiment be required to act in Skirmishing Order, it may proceed as directed in Regulations, Part V.

3. Whenever Light Troops are thrown out to protect Movements, whether they consist of the same Corps, or are composed of portions of several, they are generally placed under the superintendence of a Field Officer to command the whole. Should no Field Officer be appointed, the senior Officer will assume the command, and direct the Movements. The post of the Commander, when the Line is extensive, is properly near the Centre Support; but of course he will go wherever his presence may appear most necessary. He will always have a Bugler attached to him, and likewise a non-commissioned Officer, or a steady man or two to convey orders. When covering the formations of Troops, whether in Lines or in Columns, he will endeavour to understand the nature of the intended Movements, so as to conform to them, and be certain of co-operating with exactness, as far as circumstances will permit; and will at the same time take care that the Supports and Skirmishers preserve their relative positions and distances; because in such cases the Movements of the Skirmishers will be regulated by those which the Troops in Rear execute, to which they will conform, so as always to be interposed between them and the Enemy, to keep him at a distance, and prevent his annoying them. Therefore to ensure accuracy, when several Companies are acting as Skirmishers, one of them with its Support will be named as the regulating one, to which the others will conform. In general, the directing point of the Line of Skirmishers will be opposite to the centre of the regulating Battalion, unless otherwise ordered. Hence when any new formation or change of position takes place, they must rapidly change also so as to correspond with the new order of the body they cover. But in acting on the field, the Movements of Light Troops and

Skirmishers must frequently depend upon circumstances and situation, and will therefore be determined upon on the moment by the intelligence of the Officer in command on the spot; on which occasions the Officers must use their judgment in taking advantage of localities of ground, or errors of the Enemy, &c., which may render it necessary to adopt a different disposition from that which may have been ordered with regard to the whole or a part of the Line of Skirmishers. Thus, in extensive Movements or changes of Front, should Columns or Lines be threatened by the appearance of the Enemy's Light Troops at an unexpected point, the Skirmishers, if not engaged, may rapidly take the shortest way to where the Enemy appears; or the Supports may move to cover such point or new front, should not a fresh Line have been formed from the Reserve.

4. The following example proves how an enterprizing Officer may act on emergencies with effect, and also how hostile Columns, advancing to Attack, may be taken in Flank.

At the Battle of Sabugal, 3rd April, 1811, Colonel Beckwith, commanding a Brigade of the Light Division, was ordered, through the mistake of a Staff Officer, to Attack, at the wrong point and too soon, the Corps of General Regnier, 12,000 strong, supported by Cavalry and Artillery. He passed the Coa, there deep and rapid, ascended a steep and wooded hill, drove back a strong body of the Enemy formed to receive him, on which Regnier (who had bivouacked on the low ground) ordered up an overwhelming force, which was advancing with rapidity, and had gained the lower range, when Captain Hopkins of the 43rd Light Infantry Regiment, with a Flank Company, run out to the Right, and with admirable presence of mind seized a small eminence, commanding the ascent up which the Enemy's Column was advancing to attack the Right. He opened such a sharp fire (on their Flank) that it threw them into confusion. He then charged, and obliged this Column to recoil in disorder.[*]

5. When Skirmishers are thrown out to cover the formation of a body of Troops, they will seize the most favourable ground which lies in the direction of the Enemy, provided it is not too far off. They, however, should always be advanced at such distance as will effectually secure the Troops from the fire of the Enemy. In defensive positions they will extend

---

[*] See Napier, vol. iii., book xii., p. 489.

and line fences, broken ground, &c., and if houses, garden-walls, &c., present themselves at different points, part will occupy them so as to be enabled to take the Enemy in Flank, should he endeavour to penetrate between them. When the Troops in offensive operations advance to Attack, the Skirmishers will be pushed on to drive in those of the Enemy, and will seize such strong ground as may be favourable for the Troops when they arrive to deploy, &c., and which they will hold until the Troops are ready to engage.

6. The chain of Skirmishers will always be prolonged beyond the Flanks of the Line or Body which they are to protect, and should rest their own Flanks on such strong points as may offer. And should any strong ground lie beyond the Flanks which the Enemy might occupy to our disadvantage, a Detachment ought to be sent to take Post there.

7. When Troops advance to Attack, Parties may be sent to move on the most exposed Flank of the Columns or Lines to protect them, and counteract any attempt of the Enemy to disturb them. Likewise when the army is in Position, when the Enemy advances, Parties may be sent to endeavour to gain his Flanks, if in strong positions, by lying behind trees, rocks, &c., on the brows of heights or otherwise, by creeping behind fences, &c. But they ought not to open their Fire until he commences his Attack. For this purpose Riflemen may especially be employed; but they must at the same time act with caution, that they themselves do not get committed and cut off by similar Parties of the Enemy.

FLANKING PARTIES.

8. Skirmishers Covering Troops halted in Position will be recalled when the Enemy presses closely on to Attack. They must retire with the utmost celerity, so as to clear the front as soon as possible; or if driven in by superior numbers, they will in either case make the best of their way through the intervals of Battalions, and Flanks to the Rear, and then form up, and afterwards move to such Points as may be directed. If Covering Troops advancing to make an Attack, the Skirmishers will be ordered, as soon as these are prepared to Engage, to clear the Front, and will proceed in the same manner; or on some occasions may lie down until the Troops have passed over them, when they will Close and form up, moving to the Rear.

RECALLED.

9. During all such operations on a large scale, the Skir-

mishers Covering the Troops when a serious Attack of Cavalry is apprehended, will generally be recalled. Such portions as can occupy any garden, farm yards, or other Cover that may offer in Front, will do so to take the Cavalry in Flank. In such cases, when the Skirmishers are recalled, or are driven in, they will always make to the Rear of the Squares formed by the Main Body, and form in the Rear of them, in such manner as may be ordered. They will take care in retiring to make for the intervals between Squares, so as never to run across the Front of a Square or Column. However, should Cavalry suddenly appear, and render it necessary for the Skirmishers and Supports to form Squares, they will proceed as directed in Section XIII. But if Cover is at hand, they will avail themselves of it.

10. When at Field Exercise, Light Troops Cover a Brigade, Division, or other Body, the Skirmishers will conduct themselves on the same principles as required to do when Covering a Battalion; because on most occasions the same rules are more or less applicable. The Corps or Companies of Light Infantry when required to act, will usually take Post at due distance in Front of the Body they are to Cover, and then be thrown into Skirmishing Order, as the case may require. See Section VIII.

11. When Skirmishers are thrown out to mask the movement of a Column, or contiguous ones, proceeding to Attack, they will in general extend so as to Cover the ground which the Line will occupy when deployed, unless otherwise directed. But if, from an apprehension of Cavalry or other cause, the Skirmishers are only progressively to Cover a Column deploying, or the Flanks of one which opens out to wheeling distance, they will in this case gradually extend as the formation proceeds.

12. When Skirmishers are thrown out to Cover the immediate advance of a Line or Column, the chain, after extending, will continue to move on, until further orders are sounded.

13. If thrown out to cover a deployment, they will, unless otherwise directed, in general after extending, occupy the best Cover at due distance in Front, to protect the formation, and wait for further orders, whether to fire, advance, retire, &c.

14. When Skirmishers are Covering a Battalion, or Brigade, &c., which if in Columns Face to a Flank, or if in Line, breaks into open Column, and takes ground to a Flank;

the Skirmishers will Face to the same Flank. Or if the Columns or Line gain ground by Echellon, or the Diagonal March, the Skirmishers will make a half-turn in the same direction; and in either of such cases will proceed gaining ground; and if firing at the time, will continue doing so on the march. When such bodies resume a parallel Position or Front, the chain will do so likewise.

15. Should a Line form close or quarter distance Column Skirmishers will in general, unless otherwise directed, remain in their extended position to await further movements, in case the Columns after advancing or retiring should be required to deploy again.

16. When Skirmishers are thrown out to Cover a Battalion in Line at Field Exercise, which changes Front to a Flank; the Company destined for this Duty, which in general will be a Flank one, will wheel up either the quarter circle, or only such number of paces as may be required, so as to bring it parallel to the intended new direction, according as the new line is to be direct or oblique to the old one. Suppose the change is to be made to the Right, and the left brought forward direct, or at right angles with the old Line; the Right Company will wheel up to the Right, and Advance extending from its Right. If the Line is to be oblique, it will only wheel up the required number of paces, then Advance extending, &c.

BATTALION CHANGING FRONT.

17. If the change is to the Left, the Left Company will act in a similar manner, wheeling up to the Left, &c.

18. If the change is to the Rear, and if the Left is to be thrown back, the Left Company may wheel on its Centre to the Left, either the quarter circle, or only such number of paces as may be required; according as the new Line is to be direct or oblique, and then Advance extending from its Right.

19. If the Right is to be thrown back, the Right Company will wheel on its Centre to the Right, &c.

20. If the change be made on a Central Company, the Flank Company on the Wing to be thrown back, will in general be ordered to Cover it, and will wheel up on its Outward Flank the angle required, according as the Line is to be direct or oblique, and will then Advance extending from such named File as will entirely Cover the new Front.

21. But when Skirmishers are already out Covering movements, and if during which the Battalion or Brigade is to

change Front either oblique or direct by a Wing being thrown backward and forward, the following may be done. Suppose a Line is to change Front to the Left in an oblique direction, by the Right being brought forward, the chain of Skirmishers will bring right shoulders forward, the Left Support will bring right shoulders forward a few paces, so as to place it parallel to the new Line, and will then Advance extending from its Right, to prolong the chain so as to Cover the Left Flank of the Line or new Front. While one or two Subdivisions of the Right Flank of the old chain may be withdrawn, and will form the Right Support. The other Support may move to the Left, to replace the Left one which had extended.

22. If the Right of a Line is to be thrown back, the Right Flank of the chain of Skirmishers will do so likewise (either at once, or if firing, by left shoulders forward of each alternate rank while retiring). The Right Support will wheel up and Advance, extending to join the Skirmishers, and prolong the chain to the Right; and a portion of Skirmishers may be recalled from the Left Flanks to form the Left Support, while the former one will move to form the Right one.

23. In most cases, when Skirmishers are to be thrown out to Cover the changes of Front of a Battalion or large Body, if the change intended be a forward one, the Skirmishers will extend upon a Line parallel to that which will be the new front. But if the change of position be to the Rear, they may extend parallel to the Line previous to the change being made; and afterwards gradually conform to the movement so as to protect the Troops during the change, by bringing right or left shoulders forward, as the case may be.

24. In some cases, during movements or changes of Front which a Brigade or Division performs within itself, the Skirmishers may, previously to the change commencing, be thrown out in the new direction, or may be sent in close order to take a Central or Flank Position as the case may be; ready to Cover in such direction as may be required. Because, if the Line of Skirmishers were to perform every movement of a Brigade or larger Body, on the great arc of a circle, the men would be exhausted, and much time lost, which would defeat the object they are intended for.

25. When Skirmishers are thrown out as a Feint to mask the movements of a Brigade or other Body which is to change position to a Flank, a new Line of Skirmishers also may be

thrown out in the new direction to protect the movement; while the old Skirmishers will remain in their extended position, until the formation is completed; when they will close to the Flank nearest the new Line, and either rejoin the Brigade, or will become the Supports to the new Line as may be ordered.

26. Suppose the Right Wing of an Army is threatened with a Flank Attack, and that a Division or a Brigade has to be thrown back to meet it; a chain of Skirmishers may be thrown out to the Front, and mask the movement; while another chain will also be thrown out on the extreme Right Flank, either perpendicular or oblique to it according as the new Front is intended to be. This Line will Cover the Corps during the movement, whether it is conducted by Echellon, Filing of Division, or in quarter distance Columns. The old Line of Skirmishers will be withdrawn when the formation is completed, and the other Line may gradually retire, following the movement, and occupy a position at due distance in Front of the new Line to protect it until recalled or fresh orders given. Had the Spanish Army done this when Soult first threatened their Right Flank at Albuera, they no doubt would have accomplished the required movement under protection of Light Troops so disposed in better order than they did, and obviated much confusion which afterwards occurred.

27. Suppose Skirmishers are covering an extensive Line, or a Brigade in contiguous Columns which has to change Front to a Flank, either direct or oblique to meet a Flank Attack; and that it is wished that the original Skirmishers should keep up their fire and continue to Cover the new Front, it may be done as follows:—If to Right, the Right Wing of the Skirmishers will be wheeled up into the required direction. Its Supports will also wheel, move up, and extend to prolong the chain to the Right, and complete the new Line; while the other Wing of the Skirmishers will Close to its Right, and with its Supports wheel up and proceed to form Supports in their respective places in Rear of the new Line. See Section XII.

28. When Skirmishers are thrown out to mask a Retreat they will in general extend at a short distance in Front of the Line; or, if ordered, may, after moving out a certain given number of paces, extend on their own ground; in either case will remain halted, and await further orders, either to fire, retire, &c.

29. When a Line retires by alternate Wings, if the Skirmishers are not required to Cover the whole, a portion may Cover each Wing while retiring, or move on its Outward Flank ready to act when required.

30. When Troops Retreat by alternate Divisions or Battalions, the Skirmishers, if not required to Cover the whole, may be divided in the intervals of the halted Line; and when it retires and passes through the second Line, they will remain and occupy the intervals as before, &c.

SECTION XV.

Light Troops in General Action.

1. LIGHT Troops are of essential service in all general actions, and when properly employed, may render various and important services; because although skirmishing may seldom lead to deciding an engagement, it is particularly adapted to prepare the opening of great battles, by protecting the preparatory movements, or masking intended ones; also for protracting the commencement until the Reserves are ready, or until expected reinforcements have time to come up. Their special duty is to clear the way when troops advance to attack, by covering their front and flanks; or in defensive positions to keep the enemy's Skirmishers in check from annoying our Troops, and to harass his columns while coming on; to follow up the pursuit after victory, or to cover the retreat in case of defeat, &c.

2. In offensive operations, when an attack is to be made, Light Troops will be thrown out to Cover the lines or columns in their advance against the enemy; especially if the ground to be passed over is broken, having strong points, or partly wooded, or partly fields of standing corn, &c., where it may be supposed he may have Light Troops posted. Parties will likewise be sent to move on the outward or most exposed flanks, and protect them during the movement. By this means the enemy's piquets or Skirmishers will be vigorously attacked and driven in, so as to prevent our Troops from being much annoyed or obstructed. Should the enemy's Skirmishers not give ground, ours must be reinforced, and an endeavour made to outflank them by prolonging our line. When the Troops are established on

COVERING AN ATTACK.

the points of attack, and the decisive moment arrives for them to engage, whether by opening their fire or preparing to charge, the Light Troops will be withdrawn, and the front cleared with all celerity.

3. On such occasions Skirmishers may be employed to connect the movements and assist contiguous columns of attack advancing to storm a position. In these cases they may either be posted in the interval between the columns, and on the outward flanks, or they may cover the whole front. They must keep up a lively fire, to mislead the enemy and induce him to open his fire against the whole extent of the attacking front; because if two or three columns were to advance without Skirmishers, the enemy in position would give a concentrated fire on them, and thus occasion a serious loss; Skirmishers may therefore be considered as conductors of the enemy's fire. As soon as the columns rush on to attack, or have deployed to fire, the Skirmishers will retire. As an instance of the above, and to show what active Light Troops may do in a general action, the following may be mentioned: The French at Waterloo made their *Infantry* attacks in columns covered with swarms of Skirmishers. These were so intrepidly daring, that creeping on to the edge of the crest of the position, they galled our columns, lines, and squares to a very serious degree, allowing no respite. Their balls like bees kept flying about, stinging in every direction. Being supported by the columns, and a great superiority in Cavalry, our Skirmishers were restrained from keeping these active " Tirailleurs" in check, so that our foreign allies frequently faltered, and were with much difficulty induced to maintain their ground, as may be gathered from Siborne's sketch in elucidation of his model of Waterloo, wherein he states that " the first attack of the Imperial Guards was supported on its right with great effect by a column of D'Erlon's Corps, which boldly advanced against Alten's Hanoverian Division, under cover of a cloud of daring Skirmishers, who, sustained by some guns, pushed forward so close to the crest of our position, in rear of La Haie Sainte, and maintained so brisk a fire against the squares of Count Kielsmansegge's Brigade, that the whole Division was shaken by the furious onset, and the greater part compelled to give way. To the right of these the French Skirmishers, who extended from D'Erlon's Column to that of the Imperial Guards, to connect the attack, were pressing very closely this part of the Line. The Prince

of Orange gallantly led on two Nassau Battalions against them, but on his falling wounded, they gave way, and communicating their disorder to the remainder of the Brigade supporting them, they were rallied again with much difficulty; while at the same time five Battalions of Brunswickers, who were moved up to the right of the Nassau Troops, were assailed with such suddenness and violence by the French Skirmishers, that before they were halted and formed up, they were thrown into great disorder, and the whole Brigade gave way before the brisk and obstinate fire kept up by these Skirmishers, and were only rallied by the Duke of Wellington in person riding up to them, who succeeded in restoring order; while on the other hand it is singular to relate that the very Column of Imperial Guards which these intrepid Skirmishers covered, and prepared the way for, with such audacity, after its having arrived within 50 paces of our Line, was completely overthrown with comparative facility by the British Guards; these springing from the ground where they were lying down, poured in a destructive fire, and, with stern front, steadily advanced to the charge, which these veteran and renowned Troops declined to meet. The massive Column gave way, broke in disorder, and fled with precipitation."

4. In Defensive positions, when an Army is posted, waiting an Attack, Light Troops are thrown forward to protect the Front, by occupying such cover as the ground may afford, in order to keep the Enemy's Skirmishers in check, and prevent their annoying the Troops in position. When the Enemy pushes on in Columns to Attack, covered with Skirmishers, our Light Troops will make the most vigorous resistance, and endeavour to drive back his Skirmishers in confusion, and also to harass his Columns. To further this purpose, when cover admits, parties will be sent to gain his Outward Flank, by creeping behind fences, banks, &c, but will not open their fire until the moment he is about to commence the Attack; *e. g.* at Vimeira, the French, who came on in masses in a rapid and daring attitude, were taken in Flank by a party of the 2nd Battalion, 43rd Regiment, who stole on behind a wall, and opened their fire just as the 50th Regiment so gallantly charged them in Front. The success was complete; they were speedily broken, and retired in the greatest confusion.

5. When there are any strong points in front of a position, particularly if near one or both Flanks, such as rocks, knolls, farm-houses, garden-walls, brooks with high banks, commanding the opposite side, &c., these should be occupied by Light Troops, supported, when occasions admit, by Artillery, in order to arrest the Enemy or save our Lines from being brought into fire, until the enemy is shaken by the fire of the Light Troops and Artillery, and the impulse of his Attack enfeebled; so that the Troops may repulse him without difficulty, or proceed to make a decided attack upon him. As an illustration of this, I may give an extract from a recent French author. Expressing his opinion on the subject, he says :—" What would become of a Column of Attack if riddled " with their (the Light Troop's) fire? in what situation would " such a Column find itself when, in a short time, one-third of " its numbers was laid low in the dust? What could it do? " Before having exchanged a shot it would be disorganised, "and would be broken before it commenced its deployment. " It would in fine be in the same situation in which the Imperial " Guard found itself at Waterloo, when the Brunswick Light " Troops, posted under cover, fired at point blank, and the " bravest Troops in the world were overthrown by the most " miserable people in the Enemy's Army."

6. The necessity of Flanking parties to protect Troops advancing to attack, whether in Line or in Column, is obvious. The want of this precaution occasioned the disaster to three Regiments at Albuera; where, owing to there being no party of observation on their Right Flank, a Corps of Lancers pushed past, turned the Flank, gained their Rear, dashed through, and occasioned a most severe loss at a critical moment of the Action, which was only retrieved by the steady coolness and bravery of the Troops brought up to replace them.

7. In the Attack of the second Column of the Imperial Guards at Waterloo, the Column having advanced with its Left Flank exposed, a Battalion of the Rifle Brigade, and the 52nd Light Infantry, rapidly bringing their Right Shoulders forward, took it in Flank, while at the same time it was vigorously assailed in front by a destructive fire from the British Guards. It consequently was disorganised before it could deploy; and being thrown into confusion, gave way before the charge of the British Light Infantry Brigade.

8. Should the Enemy advance without being covered with Light Troops, the Skirmishers, particularly if under cover, should not be overawed, and ought to be reinforced, in order to render their fire more destructive, which may oblige the Enemy to Halt to return it, and by this means throw away his ammunition, and waste his fire against Troops extended and protected by cover. In such cases these may occasion great loss to the Enemy, and if his Advance is up strong hilly ground, or through passes, defiles, &c. they may succeed in repelling the Attack, particularly when they can take him in Flank, by parties lining the brows of heights or upper edges of ravines.

9. Example.—At the battle of Busaco, one of the French columns attempted to force its way up the mountain unprotected by Flanking Parties. The Light Company of the 45th Regiment under Captain Lightfoot, having got into a well selected position on its Flank, opened so destructive a fire, that the Enemy's Column was much shattered and its progress checked.

10. And to show how a skilful adversary on such occasions may take advantage of the smallest deviation from well arranged dispositions, the following may be quoted.

At the battle of Roliça, the first action fought in the Peninsula under the Duke of Wellington, the 29th Regiment, supported by the 9th Regiment, stormed the commanding heights of Roliça. During their advance up a steep and rugged ravine they were much galled by the Enemy's Sharpshooters posted on the brows of the heights on either Flank. Although this occasioned much loss, it did not however impede their progress; they succeeded in forcing the defile and gained the crest of the position, from which the Enemy, after a desperate struggle, were triumphantly driven. It was intended that the Column should have advanced under the protection of General Fane's Riflemen on the Left, and some Light Companies on the Right, but the former were at some distance, and the latter had gone rather too much to the right; the Enemy, taking advantage of this circumstance and being favoured by the nature of the ground, pushed in parties of Tirailleurs to harass its flanks.

11. It often happens in most battles that parts of both armies are posted in broken ground opposite to one another, without their approaching each other; while the Light Troops

are sent forward either with or without cannon to cover the front, in order to mask any proposed movement that may be intended, or to prepare or to maintain the action until the required moment of attack or defence arrives.

12. Light Troops may be advantageously employed when it is desirable to attain an object by developing the smallest possible force. For example, in an extensive position, where there may be a range of strong or inaccessible ground, which renders the posting of columns or lines unnecessary or their movements difficult, a Commander may be enabled to occupy a considerable extent of ground by a small body of men, and even may mislead the Enemy so as to induce him to open a fire of Artillery, and thus spend ineffectually his ammunition, while the Commander will have more available troops at his disposal. In like manner when similar strong ground occurs between different points of a position, or between different points of an attack, such places may be occupied by Light Troops, a chain of Skirmishers being established to connect them, and keep up the Line of Action unbroken. This may likewise prevent an Enemy's Column from attempting to penetrate at such points.

13. When Light Troops are covering the front of Lines or Columns in general actions, which are threatened with an attack of Cavalry, they will be recalled when the troops form squares; because Skirmishers and Supports forming in small bodies in Front, could not of themselves offer any effectual resistance so as to arrest the progress of a mass of horsemen, and would obstruct the operations of the several Squares in their Rear. However, should any very favourable strong cover be at hand, a portion of them may occupy it so as to take the Enemy in Flank. In some cases Skirmishers may be placed between the squares and on the outward flanks, to keep up a lively fire on the advancing Cavalry, but on their approach they must be ready to close rapidly to Right and Left, in Rear of the nearest Squares, ready to dash out when the Enemy turns to retire.

14. When Troops are in position, particularly when the Light Troops are engaged in Front, they may be made to lie down, or so placed behind the crest of heights, knolls, or under cover of hollow ways, fences, &c., as to be ready to spring up when required.

15. When the attack of an Enemy on any single point of a position is repulsed, and should his Columns or Lines retire

even in disorder, caution ought to be used in sending Light Troops in pursuit. These, in the impulse of the moment, might be carried too far, while the Enemy's Reserves would undoubtedly interpose to cover and draw off the beaten forces, and would with their Artillery, also open a destructive fire, which might inflict a serious loss upon the Skirmishers in regaining their own position. This should only be attempted when the strong nature of the ground warrants such a proceeding.

PURSUING THE ENEMY.

16. But when the Enemy decidedly gives way (should there be no Cavalry at hand to follow up the acquired advantage), Light Troops should instantly be sent out in pursuit, because at such a moment something dashing may be ventured. But these ought to be well supported, so as to have a sufficient rallying point if necessary, or should the Enemy's Cavalry appear. The most important point in a pursuit is chiefly to endeavour to gain sooner than the Enemy the nearest strong ground in the line of their retreat, to prevent their rallying, or to render it useless to them. The object is best attained by Flank Attacks, for which purpose parties may be sent to gain rapidly on the Flanks of the flying Enemy, while the Skirmishers press his Rear. But it ought to be impressed on Light Troops never to follow the Enemy wildly, or ever to advance to too great a distance from their Reserves. Owing to want of attention to this, much loss was frequently sustained by the Prussian and Austrian Light Troops, in the commencement of the last campaigns in Germany in 1813. They must recollect that merely running after the Enemy will not occasion loss to him, but that a well directed fire is required. In such cases, when Skirmishers have driven an Enemy from a good position, they will occupy it, and continue their fire as long as he is within reach. And when their fire is no longer effectual, they will move on again and commence another attack, always threatening his Flanks when practicable.

17. In the event of a defeat, or when Troops are withdrawn, and obliged to fall back, Light Troops will be thrown out to detain the Enemy, and cover the Retreat, should Cavalry not be employed. At all events they will generally be required to act in conjunction with them.

COVERING A RETREAT.

18. On such occasions, when acting alone, they will occupy the best strong ground which the nature of the country affords, and will hold if possible until the Lines or Columns

they cover have gained a new position. Or if the Retreat is to be continued until they have had time to gain a sufficient distance to the Rear, they will then retire from post to post in succession, but in no case will remain too long, so as to be too far separated from the main body.

19. This operation is best done by alternate Lines of Skirmishers and Supports, relieving one another. The nature of the ground, however, and the direction of the retiring Column, will in general determine the disposition of the Skirmishers. Should the Enemy's Skirmishers push on too boldly, a party may be placed in ambush, when a favourable situation occurs, which may give them a check and make them more cautious. The absolute safety of Columns or other bodies in Retreat, on many occasions depend on the skill and bravery of the Light Troops; under the protection of the latter, if they are experienced and well commanded, such bodies can, if broken, rally, recover their order, make Front again, march off, and safely gain a new position. On such occasions, the great value of Light Troops is proved, and their Commanders may acquire great credit and renown. Should the Enemy's Cavalry follow up, the Skirmishers must be on the alert, constantly prepared to receive them, and to keep pretty close to their Reserves.

SECTION XVI.

Covering Artillery.

1. WHEN Light Troops are attached to Artillery for its protection, the Commander lies under great responsibility, being, as far as his province goes, answerable for the security or saving of the guns. In case of danger, he must defend them to the utmost, and stand or fall by them, so as never to abandon the Artillery, as long as there exists the smallest possible chance of resisting even an overwhelming force with any effect. But at the same time the Artillery must not expect too much from the Infantry. They must study how to act in combination for mutual support, a perfect good understanding ought to exist between the officers of both services, who must recollect that in affairs of Outpost the utmost perseverance and capability of overcoming every obstacle or difficulty, whether moral or physical, is absolutely requisite.

2. When Artillery act with Light Troops in affairs of Outpost, the Artillery is the strength and main stay, whether acting defensively or offensively. In conjunction with it the Infantry can venture to act with greater boldness in advancing, and resist longer in a retreat; and if broken or defensible ground occurs when retiring, the Artillery can remain longer by the Light Troops than under other circumstances. The Artillery ought never to be posted on the outward Flank of a Line, unless such rests upon an inaccessible point on which it can be placed. It ought always to have a party of Light Infantry or Cavalry beyond its outward Flank to protect it from surprise.

3. Artillery ought at all times to have a special party appointed to accompany it for its particular defence, whether in action or on the march, in a country occupied by the enemy.

4. When Light Troops are to cover Artillery engaged, the line of Skirmishers will be disposed on one or both flanks of the guns, according as cover may offer, or circumstances require, and generally on a level with them. In some cases when the ground is favourable and admits of it, a portion of the Skirmishers may be posted in front of the guns: as for instance, when they are posted on a height, or a bank having steep sloping ground or a hollow in front, when they can fire over the Skirmishers without danger. It is seldom advisable to place Skirmishers between the guns, as they would be in the way of the Artillerymen. But on all occasions a portion of the Skirmishers ought to endeavour to gain such position as may enable them to gall the flank of the enemy, and disable the Artillerymen and horses, and also his covering parties. The supports and reserve, and indeed the whole of the Light Troops which are not absolutely required to engage should always be posted on one or both flanks of the guns, and never directly in rear of them, because Artillery always draws fire upon itself. Hence such Troops would be exposed to a destructive fire without necessity.

5. When the guns advance if the Skirmishers are not required to cover the front, they will move parallel with them. When guns are to retire the Skirmishers should be previously posted in a defensive position until the guns have gained some distance, when they will follow, continuing the Retreat. On these occasions, whether Advancing or Retreating, a party should always move on the most exposed Flank of the guns, or both if necessary.

6. Should Cavalry make an Attack and the Artillery form Square, the Skirmishers and Supports will form Square in the best position they can, or part may gain the best cover to take the Dragoons in Flank. The immediate escort of the guns will occupy the intervals and angles of the Artillery Square, while part by mounting the tumbrils or creeping under the guns may fire through the wheels. The Reserve will move in Square to such point as may best interpose between the whole and the Cavalry.

7. When Infantry are required to act with Artillery in a street or in defending a pass, hollow road, &c. where the breadth admits, the following method might be adopted. Suppose a gun is placed to fire down a street, supported by a Company of Light Infantry. The Company in close order may be faced outwards by subdivisions, and each moved up

in File on both Flanks of the gun, nearly on a level with its muzzle, both remaining in File faced towards the Enemy. On the order to "Fire" the two front men of each subdivision will fire, and instantly, turning to right and left, will retire along the Flanks to the rear to load. The two next men immediately take a step in advance and fire, then retire as before, &c. In this manner by the remaining Files closing up to the front at every shot, the firing may be continued without almost a moment's intermission for any length of time, because, the men who have fired always form in succession in File again in the rear.

8. And if the men level obliquely, a cross fire is given of so rapid and destructive a nature, in conjunction with the Artillery, that no ordinary Troops, much less a mob, could withstand it; and which would protect a gun from being dashed upon during the interval of loading it. When the street or defile is of sufficient breadth two guns may be employed with increased advantage.

9. Light Troops are of essential service in attacking Artillery. Skirmishers by taking advantage of cover under most circumstances have little to apprehend from Artillery. The fire of cannon even with grape shot is of little avail against men in extended order, who act with common judgment; and even in open ground, if they proceed with proper tact, their loss will not be considerable. Good Skirmishers may be considered the mortal enemies to Artillery: they, from knowing how to take due advantage of cover, may gain suitable positions from whence by a well directed fire they may pick off the men employed at the guns, and destroy the horses even before the cannon could be directed to tell with effect upon themselves. Indeed instances are not wanting when they have obliged batteries of guns to retire. And now since the shell or explosive rifle balls have been introduced the power of doing more serious injury to the Enemy's Artillery is much increased by affording a ready means of exploding the ammunition tumbrils, as has been practically proved by the French at Algiers, and the recent campaign in Africa. They would no doubt be of infinite service when a convoy of Artillery and ammunition waggons is to be attacked and intercepted during its passage through a hilly country, having woods, passes, or defiles where Riflemen or even the peasantry of an invaded country might gain favourable positions and cover on the

Flanks, to open their fire when the attack in front or rear commenced. It seems advisable that Light Troops should on all occasions, when there is a probability of their acting against Artillery, be furnished with a certain number of these balls.

10. In the combined operations of Light Infantry with Artillery and Cavalry, it requires the most intimate and well arranged combinations, particularly in affairs of Outpost, to produce or ensure effective results. In general the Infantry are principally engaged, the Cavalry have to protect the two other Arms and to ensure and follow up the victory. The combination with Artillery gives energy in such operations. It is destined to open the action and engage the Enemy at a distance, to facilitate the defence of positions, so as under general circumstances to render the maintaining of them possible, and lastly to cover and secure a safe retreat. But in these affairs Artillery can only be considered as an assistant weapon, for it contributes less to offensive than defensive measures. And this principle ought to be considered in employing it. Hence in general it ought to be with the main body occupied in the defences, where the principal stand is made. These three Arms when acting in conjunction or Outpost affair must acquire a tactical readiness mutually to support one another as circumstances require.

Cavalry and Horse Artillery.

1. As in affairs of Outpost all Arms—Light Infantry, Light Cavalry, and Light Artillery—are more or less employed, either all in conjunction, or the Infantry with one or other of these Arms, or the Cavalry and Artillery alone, it may not be unappropriate to note here the powerful effect these two latter bodies may have when brought to act in conjunction.

2. In an open country which admits of the employment of both Arms, particularly if there are extensive plains, they ought on all occasions to be inseparable, and to act in conjunction, and not be considered as merely occasional auxiliaries to each other, as was formerly but too much the case. Indeed it has happened not unfrequently that owing to the want of this constant union and of sufficient confidence in each other's mutual prowess, brilliant opportunities may have been lost.

3. " Good Infantry even in a plain will generally baffle all

"the efforts of Cavalry alone. But the steadiest Infantry in
"such a situation can hardly withstand the joint efforts of both
"Cavalry and Artillery combined, if skilfully directed. The
"very formation of Square or Column, which the Infantry is
"obliged to assume in order to resist the Cavalry, exposes it
"to the murderous fire of the Artillery, and becomes almost
"untenable. And even if the Infantry in such case is sup-
"ported by a portion of Cavalry, it will have but a slender
"chance of escape, because either the Cavalry must keep aloof
"to admit of the Infantry using its fire, and might thus itself be
"attacked and beaten separately; or, if it remains close to the
"Infantry, it unavoidably masks its fire, and moreover runs the
"risk of being driven in upon it, especially if the Infantry
"should be retreating in Column ready to form Square,
"covered by Cavalry alone as a Rear Guard.

4. "To make the most of this powerful combination of the
'two Arms a mutual and good understanding between them
"is indispensable. The officers in command of both services
"should make themselves acquainted with, or at least, to a
"certain extent, understand the nature of each.

5. "The Cavalry Officer ought to be able to estimate the
"capability of Artillery in all situations by acquiring a know-
"ledge of the various ranges of different calibres, the effects of
"the different projectiles, their proper application, and the
"nature of the ground best calculated for its movements, man-
"œuvres and effect, while he must always bear in mind that
"Artillery is comparatively a passive Arm, requiring constant
"support and protection, without which it cannot act with
"efficiency, particularly when employed in dashing affairs, or
"such operations as are here contemplated.

6. "The Artillery Officer ought to know what can be ex-
"pected from Cavalry in all situations.

7. "The Officers of each service will thus understand each
"other's measures, and thereby will prevent all unwarrantable
"expectation from being entertained on either part.

8. "The main body of the Cavalry will be disposed in such
"manner as the operations they are engaged in may require,
"according to the composition of the Enemy's force and the
"nature of the country. And the Artillery in such point or
"points as may be deemed most eligible according to the object
"in view whether Attack or Defence.

9. "The parties of Cavalry actually attached to the guns
"for the special protection and supporting the Artillery ought

" not to remain posted inactive in Rear of the field batteries when
" engaged; but ought to occupy Positions on either Flank,
" throwing out detached files to keep a look out; because
" it may be expected that parties of the Enemy's Cavalry will
" hover around threatening an Attack or sudden dash, while
" their Sharp-shooters stealthily approach, and pick off the
" Artillerymen at their guns. Hence all such annoyances
" ought to be swept from the front by the covering Detachment
" of Cavalry, so as to enable the Artillery to act with security
" and effect.

10. " The brilliant career of the British Horse Artillery
" during the Peninsular war was most conspicuous. The
" rapid and intrepid manner in which it rushed into action,
" hardly waiting for support, but seeming rather to lead the
" attack than to second it, the quickness and accuracy of its
" fire, always disregarding the greatest odds with which it had
" to contend in the superior calibre of the Enemy's guns,
" rendered them the admiration of the Army."*

* See Napier, Battle of Fuentes d'Onor, May, 1811; and a Correspondent of the United Service Journal.

SECTION XVII.

Light Troops acting with Cavalry.

1. ALTHOUGH partizan warfare requires the concurrence of all Arms, viz., Infantry, Cavalry, and Artillery, still the Infantry must be always considered the principal force, because it can, on most occasions, accomplish everything without the aid of the others, while they in general can do little or nothing without it. Detachments of Cavalry are rarely abandoned to act entirely by themselves, and in an enclosed country or strong ground, it is always necessary for them to have Infantry to support them; while Artillery are never left to act alone, but have always either Cavalry or Infantry, and frequently both, not only to act in conjunction with them, but to escort and protect them on all occasions. Hence it is evident Light Infantry have great influence in the success which may be obtained by the correct combination with the other two Arms. In affairs of Outpost they are always occupied, while the others are only secondary actors, although indispensible ones to obtain complete success in all enterprizes. Therefore Infantry constitutes the true force; it can supply the place of all the others, while its place can be supplied by none.

2. When Light Infantry are engaged in such affairs, particularly if acting in an open country, it is of essential service to have Cavalry to act in conjunction with them. In exposed situations, the Skirmishers, by having Cavalry in Support, can venture to act with greater boldness, and consequently may gain important advantages, as all their movements can be made with greater freedom under protection of Cavalry, on whom they can depend to have their Flanks secured, or their Advance or Retreat covered; while the Enemy may be constrained either to recall his Skirmishers, or withdraw them close to his Reserves, and his Cavalry will be even held in check. In the disposition of such combined force, when

Skirmishing in Action, it is not advisable to alternate Cavalry with Infantry unless the posts occupied by the Infantry are too distant to afford mutual support, or when an open space intervenes between them affording no cover, and where it would be dangerous to extend a Line of Infantry. But in such cases the Cavalry will be liable to suffer much loss.

3. When Cavalry and Light Troops are employed in Skirmishing, the following general principles will be adopted.

4. If the Enemy's Skirmishers are Infantry, our Infantry will always oppose them, having the Cavalry in Support. Also in a close country the Infantry will be in advance in Skirmishing order, and the Cavalry in Reserve. In a hilly country the Cavalry may be at some distance in the Rear, greater or less, according to the probability of their being employed. In plains or very open country, the Cavalry will be in front extended, when necessary to Skirmish, supported by Light Infantry Divisions in close order in their Rear.

5. When Infantry are Skirmishing, protected by Cavalry, the latter will be posted in Parties at intervals along the Rear of the Chain and its Supports, and the Reserve of Cavalry in close order about 200 paces behind the whole. This will tend to hold in check any Horse Skirmishers of the Enemy or small bodies of his Cavalry. On these occasions the unity of the Line of Skirmishers ought to be strictly preserved, and both, or at least one Flank of the Chain, ought, if possible, to rest upon some point of defence. It is dangerous for small bodies of Light Infantry to engage in open ground without such precaution, and only practicable when there is no apprehension of a serious attack of Cavalry. Therefore, when strong points cannot be had, a party of Cavalry will be posted behind each Flank of the Chain, and a File or two of Cavalry thrown out to some distance beyond the Flanks, or placed on heights near them, to keep a look out, and discover the movements of the Enemy, should he attempt to Outflank or turn the Position, or make a dash on the Chain by penetrating in Rear of a Flank. Likewise a Non-commissioned Officer, and a few Files of Cavalry ought to be placed in Rear of the centre of the Chain to be at the disposal of the Commanding Officer to convey orders, or to be sent to different points to observe the Enemy, &c. And on all occasions when Infantry are Skirmishing in woods or villages, Parties of Cavalry must keep a look out to prevent their being Outflanked or surrounded.

6. Should the Enemy likewise have Cavalry behind his Light Infantry, our Cavalry must keep a sharp look out, and narrowly watch them. On the smallest apprehension of their attacking the Chain, the Skirmishers will form in Squares with their Supports. Or if the Cavalry dash suddenly on them, Rallying Squares on their own ground as directed Section XIII. In either case our Cavalry will advance through the intervals, and charge the Enemy, while the Squares will endeavour to unite, or gain a strong position. Should the Enemy have surprised the Skirmishers before they had time to rally in any defensible shape, part of the Cavalry will rapidly attack him in Flank, to repulse him, or at least use every endeavour to arrest his progress, while the remaining part will rescue the Skirmishers as they best can.

7. Should the Enemy during a Skirmish have no Cavalry close up, some of ours may make a dash at his Skirmishers, either in their front, or by turning a Flank and penetrating in Rear, according as the ground or other circumstances may favour or warrant their doing so, which will be much facilitated by a bold advance of our Skirmishers.

8. Example. When Marshal Ney, who commanded the French Rear Guard during Massena's retreat from Portugal, took up with great skill a position in front of Redenha to cover the French Army during its passage through a dangerous defile, Sir William Erskine attacked his Right with the 52nd Regiment, the Rifle Brigade, and part of the 43rd Regiment. They carried the ascent, and after clearing the woods of his Light Troops, our Skirmishers advanced on to the open plain in front of his Line, which opened a heavy fire upon them, while at the same time a Squadron of the 3rd French Hussars, led by Colonel La Ferrier, made a sudden dash, charged our Chain of Skirmishers, and took 14 prisoners. This officer never failed to break in upon the Skirmishers in the most critical moments with his Cavalry, sometimes with a Squadron, sometimes with only a few men, but during the whole campaign he was always sure to be found in the right place, and was continually proving how much may be done even in the most rugged mountains by a small body of good Light Cavalry.*

6. When Skirmishers pass from one cover or position to another, either in Advancing or Retreating, if the Enemy's

---

* See Napier, vol. iii., book xii., p. 464.

Cavalry is very near, they will first close and move in that order, under protection of our Cavalry to the next position o attack or defence, and then extend again.

10. When Cavalry are Skirmishing in open ground, supported by Light Infantry, the latter will be posted in a suitable position in the Rear, ready to act as circumstances may require. And whether formed in Supports or in Reserve, each will be in close order, and be prepared to form in Squares at any instant. Should the ground become at parts broken, and affording points of defence, particularly if on a Flank, the Light Infantry will occupy such positions, and the Cavalry will continue to act in the open parts. If the Enemy's Cavalry is superior to ours, some Subdivisions of Light Infantry, should cover offer, may be brought up to the front, and open their fire, who, from being in close order, can speedily assume a posture of defence, should the Enemy's Cavalry push on.

11. Should the Enemy throw out Infantry Skirmishers against our Cavalry while Skirmishing, the Cavalry will withdraw between the Infantry Supports, part of which will extend to cover them. Or the Infantry may be concealed in a suitable position, while the Cavalry, by a pretended retreat, may decoy the enemy and draw them under the fire of the Infantry so posted. Should the Enemy withdraw his Infantry Skirmishers, the Cavalry will move out.

12. Should Cavalry be obliged to retire on Infantry, and particularly if they rally behind the Infantry Supports or Reserve, they will re-form about 50 paces to the Rear, and move so as to allow these bodies in Square to fire without hindrance on the Enemy, and be ready to dash on the moment he is repulsed or turns to retire.

13. When Light Infantry and Cavalry act in conjunction, whether supported by Artillery or not, in offensive operations, and if the Enemy to be attacked is posted in strong ground, having Cavalry in his front, our Cavalry will endeavour to drive back those of the Enemy. And if this succeeds, they will halt out of reach of the Enemy's fire, and the Light Infantry will then make such dispositions to attack as the position of the Enemy may require; this will generally be done in Skirmishing Order, covered by Artillery. While part of the Cavalry will watch the Enemy's Cavalry and protect the Skirmishers from them, another part must endeavour to threaten his Flanks, especially if he is in a village or a wood;

as when it can be done, an Enemy's position should always be endeavoured to be turned, by detaching troops to develope themselves on his Flanks, while the others or main body occupy him in front; and such demonstration may cause him to retreat and evacuate such places without much loss on our part. Should the Infantry's attack fail, the Cavalry will protect them while forming again to renew it; but if the attack is successful, some Cavalry will be pushed on to increase the disorder of the Enemy and make prisoners in the pursuit, provided the ground is favourable. But if the country is wooded or enclosed, they ought not to proceed, as the broken foe would get over the fences on the road-side or other places and harass our Cavalry without danger. In such case Light Infantry must follow up the Enemy.

14. When a Position or a Defile is carried, the main body of Light Infantry will occupy it and re-form, ready to protect the Cavalry, should they have gone in pursuit, in case of their being driven back, and also to be prepared to move in Advance to continue to press the Enemy, which will be done as follows:—If the ground is open, the Cavalry will move in Advance, followed by the Infantry in Column of march. If the country is broken or enclosed, the Infantry will lead, supported by the Cavalry. In either case every endeavour will be used to overtake him and attack, should he not effect his retreat in good order. Should he gain and occupy a position, dispositions will be made to dislodge him, as above detailed, according to the nature of the ground.

15. When Light Infantry and Cavalry acting in affairs of Outpost have to stand on the defensive, and expect to be attacked by the Enemy, the Infantry will occupy such strong position as the ground affords; or any defile, bridge, wood, village, &c., that are to be defended; while the Cavalry, if the ground is favourable, or at least some portion of them, will remain as long as possible in front, and parties will be sent to watch and protect the Flanks, so as to counteract any attempts of the Enemy to turn them. Advanced parties of Infantry will be so disposed in cover as to be able to protect the Cavalry if they are pressed, who, if obliged to withdraw, will, if the ground permit, rapidly clear the Front to enable the Infantry to fire on the Enemy's Cavalry. If the Enemy deploys his Infantry to attack, ours being posted under cover will make the most determined resistance. The main body of the Cavalry will be formed in Reserve in Rear of the

position, ready to dash on the Enemy should he be repulsed, or to cover our Infantry if obliged to withdraw. The Flank parties of Cavalry will be reinforced, in order to secure the pathways, and if possible to act on the offensive on the Enemy's Flanks, or at all events to keep in check any attempt he may make on ours.

16. Should the Enemy's attack fail, and should he retire, the Troops will proceed as above directed for advancing; but if our Infantry are obliged to evacuate the position, they will retire under protection of the already formed Cavalry Reserve, who, if the ground is open, will cover the Rear. The main body of Infantry will halt in Column when clear of the position, while the Supports will unite in close order until the remaining Skirmishers are withdrawn, by each division closing on the march and joining the Supports; but if the Enemy's fire is very hot, they will retire rapidly for some distance in extended order, before closing, and then form in Column with the Supports, move on to the Rear, all continuing the retreat over the open ground, ready to form Squares covered by the Cavalry. The Infantry will hasten to gain and occupy the next favourable position of defence, and prepare to receive the Enemy anew, both Cavalry and Infantry acting as directed for the first defence.

17. But if the country is intersected, the Cavalry will retire first; and previous to the Skirmishers giving up the position, village, or defile, a fresh Line will be posted in the nearest best cover, behind it, through which the old Skirmishers will rapidly withdraw, and the Retreat continued thus by alternate Lines of Skirmishers and Supports from one covered position to another. Should the Enemy push on too boldly, opportunities of favourable ground may be taken to place an ambuscade to check his rashness. Should the ground become open, the Cavalry will halt and form in a posture of defence, while the Infantry will pass through them, and then the Retreat continued to the next position of defence where it is determined upon to make a stand.

18. Even in woods Cavalry may be useful; they may be sent to move by the bye-paths, ready to be at hand to annoy an Enemy either in his Advance or Retreat on such occasions.

19. When an Advanced Corps is feeling its way in an Enemy's country, small bodies of Light Cavalry are of great utility, when attached to the Advanced Posts of Infantry, even if the country be intersected and rough. In carrying reports

with rapidity, in keeping up the communication from post to post, in co-operating with the Light Infantry, in Patroling to the Front or Flanks, in reconnoitring and other duties of the kind, &c.

20. From the mutual support that each will render the other, if every advantage be taken of the ground, and if both Cavalry and Infantry are brought into play, as opportunities are afforded, great advantages may be obtained.

21. In countries or districts neither very hilly nor consisting of open plains, but such as admits of both Arms acting, the combination of Cavalry with the Light Troops may be employed with advantage, whether in covering the Retreat of their own Army, or employed in the pursuit of a retiring Enemy, by acting alternately according to the nature of the ground, and supporting one another as circumstances require. For instance, in a Retreat, when followed closely by the Enemy, the confidence of the Cavalry would be increased, when on approaching a defile, rocky ground, thicket, or enclosures, through which they must retire, they perceive the Light Infantry taking advantage of all defensible ground, ready to keep at bay their pursuers, and thus enabling the horsemen to move under cover of their fire to more open ground in the Rear. And on reaching a plain the Infantry would feel equally reassured in finding the Dragoons ready formed and prepared to charge on the Enemy's Cavalry should they venture to intercept them in traversing it.

22. During the Peninsular War the French frequently even intermixed Cavalry and Infantry Skirmishers, where the ground was so chequered as to permit both to act with effect. As also on searching ground, each Cavalry scout being accompanied by a Light Infantry Soldier, which enabled them conjunctly to reconnoitre the ground more effectively, to guard against ambuscades, and to cope with any strong patrols or piquets of the Enemy they might chance to stumble upon. They likewise on some occasions intermingled both Arms on outpost duty and piquets. At Fuentes d'Onor, where the two Armies lay in position close to one another for several days, the French established an extensive line of Cavalry videttes, with an Infantry soldier at the side of each. They likewise established the same system in some portions of their Outposts in front of the Lines of Torres Vedras.

FRENCH INTERMIX.

23. " The French Chasseurs-à-Cheval, during the last war,

" were most efficient; being properly armed and trained as Light
" Troops, ready to dismount and skirmish on foot when required,
" were applicable in every kind of country, attended with the
" advantage that they could be detached without depending on
" the Infantry for support. If obliged to retire before a force
" of Cavalry and Infantry, and exposed to their joint attacks,
" through a broken country, a sufficient portion can dismount
" and dispute such positions as are applicable; while, under
" protection of their fire, the mounted squadrons and the led
" horses of those engaged can retire to ground more adapted to
" Cavalry movements, where the Skirmishers having retreated,
" will hasten to their horses and remount.

24. " Their utility may be still more effective when employed
" in the pursuit of a retreating enemy. Suppose, for example,
" they come up with a rear-guard consisting of Infantry moving
" through a rugged country; a due portion of the Chasseurs
" dismount, and attack in Light Infantry order, and continue to
" harass their opponents. Should the country become open,
" and the Enemy relieve their Infantry by horsemen, the Chas-
" seurs can quickly remount and renew the pursuit as Cavalry.
" During the retreat of our Army to Corunna, our Rear-guard
" of Infantry was overtaken near Calcabello by a large body
" of Chasseurs-à-Cheval (who had preceded their own Light
" Infantry,) a portion of whom dismounting, attacked our Rifle-
" men with impetuosity, and continued warmly engaged for a
" considerable time. Indeed, these dismounted Dragoons
" proved on all occasions that they were extremely well trained
" in Light Infantry matters; and there can be no question but
" that Light Dragoons, armed and instructed as these Chasseurs
" were, are a formidable description of Light Cavalry; and
" particularly so at Outpost duty, and in advanced and rear-
" guards. It may be still in the recollection of several expe-
" rienced officers of Cavalry, that when the Cavalry of both
" armies were at a distance from their respective Infantry,
" the Chasseurs did frequently harass our troops by distant
" shots from on foot, resting their fusees against a tree or rock,
" or sometimes resting them over the saddles of their horses,
" while our Cavalry were unable to retaliate with any effect;
" because at that period the British Cavalry did not possess
" such a description of carbine as to enable them to oppose
" successfully those Chasseurs at the Outposts; and were then
" unaccustomed to that description of warfare, not having been
" properly instructed how to dismount, and to skirmish on foot

"when so attacked. This is now, however, considered an "essential part of their duty.

25. "When the Light Division, in conjunction with some "Light Cavalry, was falling back from the Azarva towards "Fuentes d'Onor, it was followed by a large force of French "Cavalry; the Light Division bivouacked in a wood between "Espeja and Fuentes, whilst the Cavalry established their "piquets in close observation of the Enemy. A repetition of "shots in the direction of our Cavalry kept the Infantry on the "*qui vive*. Shortly afterwards an officer of the 1st German "Hussars came in to request that some Riflemen might be sent "across the large plain in front to reinforce the Hussar posts, "which the French Chasseurs, by dismounting a part, and "acting as Light Infantry, continued to harass and annoy, "while the Hussars, with their short carbines, could make no "defence, and were quite at their mercy. These Chasseurs "were not only thus a cause of annoyance to our Cavalry at "Outposts, being also well instructed in all the duties of Light "Cavalry, in the same manner as Hussars, well mounted, and "their fire-arms far more formidable than the short carbines of "the latter; were consequently the better adapted of the two "for the various contingencies of the Outposts, and far superior "to Lancers, whose pistol would never cope with the fusee of "the Chasseurs; nor is the lance a weapon at all well suited for "Outpost duty.

26. "It appears essential that every man posted as a vidette "should have good fire-arms. If he is only armed with a lance "or a sabre how can he give timely warning to those in his Rear "or on his Flank that an Enemy is advancing, or how is he to "encounter the Enemy's Skirmishers.

27. "The conveying Infantry soldiers mounted en croup "behind Cavalry, may on some emergencies be resorted to as a "means of covering a portion of Light Infantry in a rapid manner "to the front; but which could never be adopted as a general "practice, and would be quite impracticable in a retreat, con- "sidering the weight which all Cavalry in the field must carry, "exclusive of forage. And if in addition to which the horse had "also to carry an Infantry soldier with musket, ammunition, "knapsack, &c., the horse would be so overburdened, and the "pace so tardily slow, that nothing would prevent them from "being overtaken."*

* Lieut. Colonel Leach's Recollections, &c.

SECTION XVIII.

Advanced Guard.

1. EVERY Army, or other body of Troops, when acting in the Field, must adopt certain means of defence for the security of its Front, Flanks, and Rear. This is accomplished by detaching parties as Advanced and Rear-guards, with Flank Patrols; also by sending out Patrols of discovery in such directions and to such distances as may be deemed necessary.

2. An Advanced Guard is a body of Troops pushed on in front of an Army or Column, principally for the following purposes:—

3. GENERAL PURPOSES OF To feel the way through a country by searching for the Enemy; to gain intelligence of his situation, designs, and numbers; to watch his motions, and reconnoitre his positions: for these purposes it must scour the whole country in its front, penetrating through woods, searching villages, defiles, &c., and report whatever discoveries may have been made.

4. To guard against surprise, to give warning of the vicinity of the Enemy, and timely notice of his approach, so as to prevent unexpected collisions with him; and, if necessary, to keep him in check and detain him, until the main body may have time to prepare and make the best dispositions either for making or repelling an attack.

5. To mask the movements and formations of the Army; and to conceal proposed designs; to fall suddenly on the Enemy's Piquets when required, or to dislodge him from a post; likewise for the purpose of seizing a post, and anticipating the Enemy in some important point.

6. To pursue the Enemy when he retreats, so as to press with vigour upon, and harass his rear, and to cut off baggage and insulated Corps of the retiring Enemy.

7. In former times wars were carried on more or less by separate armies, or different "Corps d'Armée," acting on different points; while at the same time a system of partizan warfare was established in their front; or, as occasions offered, to act in all directions, in every part of the country against the Enemy; and the front of each such Corps was only covered by a Detachment from its own force, as a particular Advanced Guard to itself. But now, since the new system of carrying on war in masses generally bearing on one point has prevailed, the old system of irregular petty warfare is not required. And this has led to the institution and adoption of self-constituted "Advanced Guards," consisting of a separate Division, or a "Corps d'Armée" composed of all Arms; whether intended as an advanced Corps of observation to watch the Enemy, or destined to form the Advanced Guard in forward movements, and the Rear-guard in retrograde ones; which has given an importance to "Petty War," which it did not formerly possess. Hitherto, when great battles began, the petty war was closed, but not so now. The Advanced Guard and its Outposts begin the strife, and from their strength, maintain it with energy until supported; and before one expects it, the battle is raging; as was frequently the case with our noble Light Division in Spain, and also with the Advanced Guard at Quatre Bras, previous to Waterloo. The advantage of this disposition consists in giving the advanced posts a greater and more united self-consistency, and creates an increased security to the Army. An enemy can neither easily overpower or outflank such an Advanced Guard. He must fight, and it has sufficient strength to resist him until the Reserves come up. Besides, an Army advancing obtains by this means a point of "appui," from whence further operations can, if necessary, proceed with vigour. But although such Advanced Guards are strong and effective bodies, still the utmost vigilance and a well-regulated system of security are at all times necessary. No relaxation should ever be permitted, even when opposed to a weak or inactive Enemy. No over-weening self-confidence should ever be indulged in. An Army who manages the Outpost duty with little or no ability, and who does not display a proper degree of vigour and alertness, will sooner or later fall by some disaster into the power of the more active Enemy; or who will at least weary and harass the troops so as to worry their lives

LARGE ADVANCED GUARDS.

out. Hence an Advanced Guard ought at all times, whether by day or night, whether halted or on the march, to be held in fighting order, so as to be ready to oppose the Enemy under every circumstance. When halted or encamped, or employed as a corps of observation, the best position for defence should be chosen (similar to that required for a field of battle) which is not commanded or liable to be turned ; so that the Flanks may be well covered and secured ; and has a free range in front for the Artillery, that the guns may cover and command the surrounding country ; while all the Outpost duties, piquets, &c. must at all times be carried on with the greatest exactitude. And when on the march, its Front and Flank parties ought to display the greatest activity and intelligence. But as an Advanced Guard cannot trust merely to those internal arrangements within itself for security, further precautionary measures by a more extended sphere of action become necessary to ensure its safety. This is accomplished by a judicious system of patrolling. For this purpose patrols of discovery are sent out in such directions to reconnoitre as may be deemed requisite.

8. The General Officer, officers, and men should be selected COMMANDERS. for their capabilities and knowledge. A Corps deficient in instruction is only an embarrassment to an Advanced Guard. The following are the leading qualifications which distinguish a good Commander of an Advanced Guard.—To spare no pains to gain information by every possible means ; and always to bear in mind that vigilance and a careful study and knowledge of localities are the main-springs of action. To attain these, he must secure the service of spies on whom he can depend. He ought to understand languages, so as to be able to interrogate deserters, postmasters, magistrates, clergymen, &c. to intercept public despatches, suspected letters, &c., and translate their contents ; to provide the most trustworthy and experienced guides, acquainted with the best roads, bye-paths, fords, &c., so that he may know the country thoroughly ; to reconnoitre a district, defiles, &c. accurately, and be able to conduct a reconnoissance with skill ; to cultivate a good understanding with the inhabitants ; to give his orders for the execution of any movement or enterprise intelligibly in a few words ; in fine, to be able to give every information to, and to answer satisfactorily every question of the General-in-Chief when required. In all the operations of war, there is none where an active and skilful officer can have more opportunities

of distinguishing himself, or proving his abilities by paving the way to victory for the army he belongs to, than the enterprising Commander of an Advanced Guard.

9. Although a Commander on this peculiar service should at all times avail himself of any errors committed by an adversary, still no dependence for success should ever be placed in the probability of the enemy being guilty of flagrant blunders. The soldiers must be taught that they cannot look for success from any other source except their own zeal, discipline, and gallantry, with a watchful alertness on Outpost duty, and from the talents of their Commander.

10. An Advanced Guard on this scale being destined to cover and protect the movements of an army, it must be capable to resist the first attack of an Enemy, so as to keep him in check, and give time to a Commander-in-Chief to assemble the different corps of his army in a given position; if it is scattered in Cantonments, or if encamped or in bivouac, to get under arms and occupy their proper posts and complete their formation for attack or defence under its protection. It ought likewise to be able to clear the way for the advance of an army by main force when necessary; and when employed in pursuit of an Enemy retreating to follow him up with unremitting vigour. For these cogent reasons, and as such a corps will necessarily, during the diversified operations of war, have to act in countries presenting every variety of features, from extensive plains to almost impracticable paths, over the highest hills and roughest mountains, it must be so constituted as to be enabled, and in a fit state to act on every sort of ground. It must therefore be composed of all Arms—Infantry, Cavalry, and Artillery. But the proportion of each of these arms must depend on circumstances, and will vary according to the nature of the country, and the composition of the opposng force of the Enemy.

COMPOSITION.

11. The employment of these different sorts of Troops, and the order of their disposition in the Advanced Guard, will likewise depend upon whether the country in which it is acting is open, woody, or mountainous, &c. In an open country or plains the Cavalry are the active force, the whole or greater part of which would lead the March, forming the van, and doing all the advanced duties of Scouts, searching Patroles, &c., followed by the Horse Artillery, and the Infantry in the Rear in support of the whole. Indeed in such a description of country, if forage is plentiful an Advanced Guard may be

chiefly composed of Cavalry supported by Artillery, and a due proportion of Light Infantry.

12. During the late Campaigns, the Russians employed whole legions of Cossacks on this service, the Austrians their excellent Corps of Hussars and Hulans, while the celebrity of Napoleon's dashing Cavalry Advanced Guards, under Murat, are a matter of history.

13. In a close or hilly country, and those covered with forests, Infantry will at all times constitute the main force; in which case the Infantry may be calculated from two-thirds to three-fourths of the whole strength; the Cavalry at about one-third, with one or more batteries of Horse Artillery, as circumstances may require. While in decidedly rugged, mountainous countries, intersected with valleys and torrents, where the roads are steep, winding along perpendicular rocks on one hand, and precipices on the other, &c., affording favourable points of defence by the Enemy, or are passable only by narrow paths or slippery sheep tracks overhung with wood, behind which ambuscades might be apprehended. It is the Light Infantry alone who act, and who must be in front, and perform all the Advanced duties.

14. In a broken country, or partially open, due proportions of Cavalry and Infantry will be necessary, so as always to have at command the sort of Troops the ground may require; so that on leaving open parts, should it become diversified, whether intersected with enclosures, canals, woods, or hilly with defiles, &c., the order of March may be changed, and the Infantry placed in advance, thus alternately doing the Advanced duties in conjunction with the Cavalry, as the country may admit.

15. In rugged countries the Horse Artillery will usually move in rear of the whole, until the doubtful ground in front and on the flanks has been duly explored and passed through. A gun or two might, however, be brought up if required, should a favourable position for that arm offer.

16. The strength of an Advanced Guard will in general be regulated in proportion to the strength of the main body it is to protect, by the object in view, and the nature of the country; but it ought always to consist of such strength as may be proportional to the supposed numbers of the Enemy, particularly if it is pushed on as a corps of observation to a considerable distance, or has to anticipate the Enemy in some important point. Hence some judgment is

STRENGTH.

required to determine the proper strength, because if too large, it would become unwieldy, and if too small, it would be liable to be overpowered.

17. When an army is advancing, an Advanced Guard of this magnitude, especially if employed as a corps of observation, may be one or more days' marches in advance of it, according to circumstances; but if engaged in the pursuit of an Enemy retreating, the distance is seldom so great.

DISTANCE IN FRONT.

18. Every Advanced Guard should be provided with a sufficient proportion of axes and entrenching tools to repair roads, remove obstructions, or to form abattis, barricades, &c. Reg. 285, No. 15.

19. It is obvious that an Advanced Guard, constituted as above, is calculated for those required to cover a large army on a grand scale of operations. When a Corps d'Armée or smaller body is employed on any detached or particular service, such body will always form, from its own force, an Advanced Guard on a scale proportional to its own strength, or as the object in view may require. It is held as a general rule that every body of Troops on the march, whether on actual service or not, or whether a whole Corps or a detachment, ought to throw out an Advanced Guard.

20. The disposition of an Advanced Guard depends whether it is merely moving on the line of March, or whether, on finding itself in presence of the Enemy, it is to be formed in Skirmishing order, either for attack or defence.

On a Line of March.

21. When an Advanced Guard is moving in the line of march, whatever kind of troops it is composed of, it will be divided into the following parts; and which will be disposed of as follows:—

DEPOSITION ON LINE OF MARCH

22. A Van, or party formed as a Patrole to feel the way.

23. Flank parties, on one or both Flanks of the Van, to examine in conjunction with it all broken ground or other cover which may occur on either side of the road, so that all these may keep a good look out to discover the Enemy, and prevent his sudden approach.

24. A Reserve, destined for the support of these bodies, and to keep up the communication with the main column.

25. And when circumstances require it, Patroles of dis-

covery, sent out to such distances in Front or Flanks as may be deemed necessary to reconnoitre and gain information.

26. When an Advanced Guard is on a large scale, a due portion of it will be formed as above, and the remainder will follow in column, ready to move up and to act as may be required.

27. In very intersected districts or in strong ground, the strength of the Van and Flank parties will be regulated according to difficulties likely to occur in searching the ground, and in the necessity of doing so with greater or less accuracy; while in plains or very open countries a small Van will be sufficient. In some cases in such localities the Flank parties may be omitted, and any detached hamlets, copses, &c., may be searched by parties of Cavalry.

28. In mountainous or close countries, forests, &c., the Van will be formed of Infantry. A few Cavalry should always be attached to them, in order to bring back intelligence, and keep up the communications either between each of its separate parties or the column in the Rear.

In some cases, when necessary, collateral Patroles may be sent out on one or both Flanks of a Van, as the case may require, to move parallel with the second section of the Van at such distance as may be requisite, or the localities permit. These may consist of a Subaltern's party, if meant merely for observation. But if resistance is expected, or if there is any apprehension of their meeting detached parties of the Enemy, one or more Companies may be employed, particularly if passing through woods or strong ground where ambuscades might be concealed. At the affair of Grijo, on the 11th of May, 1811, previous to the taking of Oporto, while our Advanced Guard was moving on, feeling for the Enemy, having an open but rugged common on the left and a wood on the right, two Companies of the 29th Regiment were ordered into the wood, which had the desired effect of driving in the French parties hovering on that Flank. They fell back on their own Advanced Guard, strongly posted on a wooded height, but which being threatened on both Flanks, viz., by a column of Portuguese on the left and another of the German Legion on the right, while our Light Infantry, supported by columns in their rear, opened a lively fire along their front, gave way, and retired in great haste on Oporto.

29. Every Advanced Guard, whether feeling for the Enemy

ENEMY'S FLANKS TO BE THREATENED. or making dispositions to attack him, must hold it as a general rule that all villages, defiles, woods, &c., which he may occupy, or which may offer him concealment, must invariably be turned, and his rear threatened by Flanking parties previous to his being felt or attacked in front. By this means the Enemy will be discovered, and probably dislodged without loss, as he will naturally fall back before his retreat is intercepted, while if attacked at once in front only, he might still retire in safety after having inflicted a serious loss on the assailants. This rule is equally applicable to Patroles of all kinds.

30. When Cavalry form the Van of an Advanced Guard, a party of Light Infantry or Rifles should be attached to it, in order to be at hand to clear any barricades of the Enemy, and remove such or other obstructions they may have raised across roads, bridges, &c., but particularly to search large farm buildings, villages, &c., while parties of the Cavalry turn the flanks by going rapidly round, and halt beyond until the Infantry have finished, and re-appear to join them on the opposite side.

31. Likewise in the passage of ravines, hollow ways, or PRECAUTIONS PASSING DEFILES. roads having plantations on one or both sides, &c., the Light Infantry will make the necessary reconnoissance, and will take post at the debouche, until the Cavalry have pushed on a Patrole to observe the front for some distance in advance, to see that all is clear previous to sending back word to the main body that such is the case. Indeed during the Peninsular war, in traversing districts of a diversified nature, the French used to employ Cavalry and Infantry conjunctly in scouring the ground in front, and in examining that on either flank, by each Cavalry Scout or Skirmisher having a Light Infantry soldier at his side, supported by a Squadron or two of Cavalry, and some Companies of Infantry, in readiness to succour them, whether assailed by Cavalry or Infantry; so that by this arrangement the ground was reconnoitred more thoroughly by the joint exertions of the Soldiers of both arms. These precautions were the more necessary to them as Guerilla parties were always on the alert, and the inhabitants decidedly hostile.

32. When near the Enemy, the main column of an Advanced Guard should never enter a wood or defile, nor ap-

proach a bridge, village, &c., until the Van has passed through and reported; but should remain halted while the Van proceeds to examine, so as to prevent any chance of falling into an ambuscade. If all is right, the Van will take up a position in front of such defile, village, &c., at such distance in advance, that in case the Enemy should make an attack, the Advanced parties would not run much risk of being driven back upon the column while in the act of debouching, but that the main reserve should have sufficient room and time to form up, prepared to receive him.

33. On the 17th June, 1815, when the Duke of Wellington fell back from Quatre Bras to take up his position at Waterloo, the French Advanced Guard, consisting of a large body of Cavalry, followed up the English army. Lord Anglesey resolved to check their pursuit, and very appropriately took an opportunity of doing so while they were debouching from the village of Genappe, when the head of their column got through, and began to open out. The Life Guards and Horse Guards (Blue), attacked them with great vigour, and drove them back on their Reserve, which were still defiling through the narrow streets, and over a bridge, so that they were unable to afford the necessary support, or prevent the defeat of their Advance.

This decided and gallant measure effectually checked all further annoyance, and our Troops continued their retreat without molestation.

34. When the Van has reported that these necessary precautions have been taken, the Column will then move forward to pass the defile, village, &c. And as an Enemy is always anxious to attack a Column during its passage through a defile, or during its debouching from one, every defile should therefore be passed with celerity, and the largest breadth of front that it will admit of.

35. In traversing a mountainous district, should there be any apprehension of ambuscades, an Advanced Guard ought to occupy, with Flank parties of Infantry, all elevated heights, particularly such as command the direction of the march. These parties should remain in these points until the column has passed, and then follow in the rear. Every gorge of ravines or wood overhanging the road should be well searched. No valley, however deep, should be neg-

PASSAGE OF DEFILES.
FLANK PARTIES TO CROWN AND MAINTAIN COMMANDING HEIGHTS.

lected, because in such countries, especially if the population is hostile, one is exposed to be attacked at all times. It is only by the greatest vigilance that attempts of the Enemy can be frustrated, when the ground is favourable for such purposes.

36. The same precautions may be adhered to during operations in forests, or in a woody country, such as North America. If the wood is not very marshy, interrupted by cedar swamps, or very broken ground, nor much impeded with underwood, a double line of Patroles may be employed, one of which, should the ground admit of it, may consist of Cavalry. The localities will decide whether these are to form inward or outward ones. If the wood be not of great extent in breadth, Patroles of Cavalry may move along the outskirts, and parties of Infantry on the Flanks of the Column, who will also act as supports to the Cavalry ones. In a mixed Advanced Guard, when a hilly or a wooded country has been passed, and on coming to open plains, the Infantry may form up on one or both sides of the road, so that the order of march may be changed, and allow the Cavalry to pass to the front, and form the Advance.

37. An Officer in command of an Advanced Guard will always receive his instructions in case of falling in with the Enemy; whether he is to attack, or whether he is to take post to watch their motions, and wait for further orders or reinforcements; or whether he is to fall back. Therefore, on coming in contact with the enemy, it will depend on previous orders and particular circumstances whether the Advanced Guard is to engage or not. But if it is left to his own discretion to act as he may deem fit or prudent, the following General Rules ought to be observed:—Never imprudently to engage without knowing the strength and description of the Enemy's force. If an attack is to be made, always to find out, if possible, the weakest point of the Enemy, and attack it with as strong a force to bear on that point as circumstances will permit. Lastly, whether one attacks, pursues, or is obliged to retreat, everything ought to be done with prudence and regularity, at the same time with due rapidity, unity, and firmness.

FALLING IN WITH THE ENEMY.

38. Should an Advanced Guard be set upon unawares by a body of the Enemy, it will offer the most determined resistance, and immediately assume an attitude of attack. It will, however, depend upon the Commander's instructions,

whether he is to engage his whole force, trusting for support, or merely to keep the Enemy in check, to gain time, and enable the Column or Army to make the necessary formations. If it falls in with the Enemy at night, the Commander should instantly attack with part of his force, keeping a Reserve, and not give the Enemy time to force him back. A resolute opposition, even with a few men, will embarrass and intimidate the Enemy for some time, until the Column can prepare to act. He should endeavour, if possible, to make some prisoners, in order to find out the strength and design of the Enemy, whether it is only an Advanced Party, Patroles, &c., or a premeditated attack.

39. When an Advanced Guard on a large scale discovers the Enemy, and has orders to attack him, the disposition for doing so will depend upon the position and formations of the Enemy, whether in close or Skirmishing order, and the nature or composition of his force.

40. An Officer commanding an Advanced Guard should endeavour to observe everything himself as much as possible, by occasionally visiting the Van or Advanced Parties; to watch the Patroles and reconnoitre the country, so as to make his own observations, and trust as little as possible to others. He must keep up incessant communication with the Commander of the Column or Army in the rear; and make frequent correct and clear written reports to him of everything that occurs.

41. These are the leading principles upon which every Advanced Guard should be conducted, whether the object be to fall suddenly on the Enemy's Piquets, to dislodge him from a post, or merely to act in observation, and give warning of his vicinity.

42. The general rules which govern one on a large scale are more or less applicable to all, however small.

Small Advanced Guards in Skirmishing Order.

43. When a body of Light Troops, whether a battalion, a wing of one, or other body, is destined to form the Van, and perform the advanced duties of a Light Division, for instance, or is to form the actual Advanced Guard of any other detached body, it will be pushed on about 500 paces in front of the main Corps or Column, or to such distance as may be necessary; and will then throw forward a Company as a Van,

SMALL ADVANCED GUARDS.

formed as directed in Her Majesty's Reg., Part V., Sect. 5, No. 3.

44. But to save time, the formation of the Van, or Advance, can be equally well accomplished, and more rapidly executed, by doing it on the march progressively in succession from the rear, as it advances. On the sound " Form Advanced Guard," the named Company steps off in double time in column of Sections, suppose right in front. After gaining about 100 paces, the formation may commence by the left File of the rear, or No. 4 Section, taking up the quick step to form the Files of communication with the main body or Column. About 100 paces further, the left subdivision, or Nos. 4 and 3 Sections, take up the quick step to become the Reserve of the Advance under the Captain, pushing on the right File as one of communication. The second Section, or No. 2, on gaining 100 paces, takes up the quick step, and forms the support, under a Subaltern, pushing on a File of communication. No. 1 Section, at 100 paces further, takes up the quick step under a Subaltern or a Sergeant, pushing on the two right Files under a Corporal 100 paces, to form the head of the Van or Advance, and one File as one of communication, and also detaching a File, or a double one, to either Flank, 100 paces from the road; these to move in line with the File of communication in rear of the head of the Van. But on ordinary occasions these two Flank Files need not move out until occasion requires. This being done, all these parties and Files will gradually correct their proper distances as prescribed in the Regulations, by taking them up from the rear, according to the distance the Van has been ordered to take in advance of the Column, and the Files of communication being always mid-distance between each body.

45. When the distance in advance of the Column is 500 paces, two Files of communication may be left by the Rear Section.

46. On service, should this Van be required to be of greater strength, so as to afford stronger Flank parties to search the ground, &c., its numbers may be increased by employing two or more Companies, and substituting Subdivisions for Sections.

47. The remaining Companies will follow in Column at about 500 or 300 paces distance, according as it may be day or night, clear or misty: sending out, if necessary, Flank parties to one or both Flanks. Should the Van give intelli-

gence of the Enemy being discovered, the Reserve will move up, ready to be formed in Skirmishing order, in the best position for attack or defence, as the case may be, because an Advanced Guard, unless in presence of the Enemy, need not be preceded by a chain of Skirmishers.

48. The arrival of the main column or body will determine further operations.

49. Should some Squadrons of Cavalry, with a few pieces of Artillery, be attached to an Advanced Guard of this kind, the dispositions of these on the march, and the employment of them in feeling for the Enemy, will be regulated according to the nature of the country, agreeable to the principle above detailed for Advanced Guards on a grand scale. And on coming in contact with the Enemy, the description of force required to act, whether Infantry or Cavalry, or both, in conjunction with the Artillery, will depend on the nature of the force of the Enemy, or whether he is in close or Skirmishing order. Some details for the disposition and management of these different arms, whether acting conjunctly or separately, under the various circumstances which may occur, will be found under the several Sections on these heads; but in all cases, whether they act offensively or defensively, parties must be thrown out on each flank to prevent their being out-flanked or turned.

50. The distance of an ordinary Advanced Guard, such as this, in front of the main body must be determined by local circumstances, such as its own strength or that of the main body; the nature of the country, whether open, plain, intersected, woody, hilly, &c.; or on the state of the weather, whether clear or hazy, and whether day or night. It should, when circumstances permit, be at such distance that all villages, defiles, &c., may be examined before the arrival of the Column, so that there may be no delay; or that, should it come in contact with the Enemy, the main body may have time to put itself in a posture of attack or defence; but it ought never to be at so great a distance from the main Column as to run any risk of being cut off.

51. When Advanced or Rear Guards are formed from the Piquets of an army, under the Field Officer of the day, the formation will be regulated as laid down in Her Majesty's Regulations, No. 3, page 284; and one-third part at least will be held in reserve, whatever number of Companies they may consist of.

Duty of Advanced Files and Flankers.

52. When feeling for the Enemy, the advanced Files and Flankers ought always to consist of parties of double or single Files at least, so that when one is examining any place, the other may look out and protect him.

53. Their special duty is carefully to examine all villages, farm-houses, enclosures, copses, &c., within their reach; but should distant objects present themselves, Patroles or extra Flank Files must be detached from the second Section, for their examination. They must proceed with due caution, and look well about them, that the Enemy may not fall upon them unawares, or cut them off.

54. They will question all persons they meet, and prevent them (especially if mounted) from proceeding towards the enemy, or conduct them to the Officer commanding the Support, as such people may prove useful guides.

55. The advanced Files will follow one another at the distance of about 50 paces, so that each can always see their comrades in front; and on coming to a bend in a road or the corner of a street, the first one will look round the angle, and halt, until relieved by the next one, who will signal to the rear that all is right, when the whole will move on. By this means the chain of communication is kept up; and should the first File be killed, the others have time to fall back and report. The Flank Files on either hand always move in a line parallel with the File of communication next the head of the Van.

ORDER OF MARCH.

56. On approaching a village which might be occupied by the Enemy, an Advanced Guard will proceed with great precaution. The Advanced parties and Reserve of the Van will be halted beyond the reach of musketry in a good position, while strong Flanking parties will move round the outskirts to threaten his rear. The Advanced parties may then move on, followed by as many additional Files in succession from the Support as may be deemed expedient. If it is ascertained that the place is not in the occupation of the Enemy, the Support and Reserve will move forward; but in every case, and especially if the Enemy has but just evacuated the village, one or two of the leading Files should always gain the belfry of the church, from which the surrounding country can be

VILLAGES.

surveyed, so as to observe and give notice if there is any appearance of the Enemy, before the Advanced Guard proceeds.

57. Should the Enemy still hold the village in force, or only have small parties remaining in it, the necessary dispositions will be made according as the case may require, to drive him out, as directed. Section xxiii.

58. The Lead or Van of an Advanced Guard must never commit itself by entering a defile, pass, or hollow way, without the heights on either side being previously occupied by Flanking parties. When the heights are thus crowned, the leading advanced party will move on by files in succession, those in Rear always keeping the preceding ones in view; while the Flank parties on either hand keep moving on, in a line with the second advanced file, until the defile is passed; after which, if there is no apprehension of the Enemy, the whole will move on in their original formation. But if the Enemy be still, (even although at some distance) in front, the Van should halt, and hold the debouche of the defile until the Reserve or main Column of the advanced Guard comes up. When the brows of a defile or pass are of considerable extent, broken, rugged, or wooded, where an Enemy might be in ambuscade, Flanking parties may be extended at 20 or 30 paces between Files, so as to form a chain across while advancing.

DEFILES.

59. In passing through a wood, Flank parties ought likewise to be extended in the same manner, exclusive of Patroles thrown to a distance on either hand. And in some cases, it may be necessary to station parties of Light Infantry at intervals on either side of the road, until the Column has passed; because in a wooded country, or one covered with forests, the greatest precaution must be used in traversing them, particularly in such a country as America, where a skilful and enterprising Enemy can readily form ambuscades. Precise information, and good guides, having a knowledge of the roads and paths, are absolutely necessary. On approaching a suspicious place, whether a defile or in a forest, or where there is reason to suppose that the Enemy may be in ambuscade, the Van will halt in a position as much concealed as possible. Scouts and additional Flankers will be sent out to explore, and carefully examine the defile or woods within a certain

WOODS.

distance of the road. Should no Enemy appear, the Van will move on; but after passing the defile, or suspicious part of a forest, it will take up a position at the head of it, to protect the Column, which will not enter the defile or move through the wood until it has received a report from the advance that the necessary precautions have been taken.

60. Swamps, large sand-pits, quarries, or other broken ground, particularly if covered with brushwood, must be carefully examined, otherwise an enemy might remain concealed within the chain, and would not only risk the Advanced Guard's being cut off, but might compromise the safety of the Column.

61. Should the Flank parties come to impassable obstacles, such as morasses, swampy bogs, deep ravines, &c., which would intervene between them and the advanced Files on the road, they must close into the division from which they were detached, so as not to leave such ground between them and the Van, lest they might be cut off; but a File should attempt to explore or to pass such obstacles, to ascertain whether it is occupied or passable or not for the Enemy.

FLANK PARTIES.

62. The Flank parties will on all occasions pass over the highest ground on the Flanks within their reach, so that they may see about them. On coming to a fence, one man will get over first to examine, before the others follow. Should perpendicular rocks, &c., interrupt their progress, they must close in to the advanced Files of the Van, ready to move out as the ground expands again.

63. On coming to a river, if it is fordable, either on one or both sides of the bridge, the Flank Files must endeavour to pass over, particularly if the opposite bank is elevated or covered with brushwood, where an Enemy might be concealed. But if it is not fordable, they will close in and pass the bridge with the advanced Files, and afterwards resume their proper places.

64. No specific rules can be laid down for the Van of an advanced Guard in every case of falling in with the Enemy: this must in general depend upon the particular circumstances of each case, and the object in view. The Officer commanding the Van will act according to the orders he has received.

65. The advanced parties, Flankers and Patroles, on discovering the Enemy, or on his first appearance, must not fire, as by doing so they would discover to him that the army was near, and prevent a chance of surprising him. They should rather fall back under cover, or hide themselves on the spot, and make the preconcerted signals to the Rear, according as the Enemy may be stationary or advancing. Thus, in either case, securing all the advantage of a surprise, they must watch his motions, and endeavour to ascertain his strength, or whether composed principally of Cavalry or Infantry, &c. A report of what is observed will then be sent by a man to the Rear. However, should the Enemy fall suddenly upon them, they must instantly fire to warn the Troops, and enable them to prepare to receive him. Or when the case admits, the Van or Advanced Party may close up, and take post as an out-piquet, with a chain of sentries in front, and so remain until the main body comes up, and dispositions made either to attack or receive the Enemy. It must be impressed upon every one on all occasions of Advanced Post duty, that nothing is more reprehensible than offensive or defensive measures being undertaken in an isolated manner. Advanced Parties and Patroles should never be suffered to engage of their own accord, if it can possibly be avoided. Reg. p. 287, 288.

DISCOVERING THE ENEMY.

66. If a weak Advanced Guard or Van is ordered or obliged to retreat, the Officer will fall back gradually to a Flank, and not directly on the Column. He will by this means avoid any confusion, should his men be obliged to make a precipitate retreat, and he will thus leave a clear stage for the operations of the Column.

FALLING BACK.

67. When the Column halts, the Advance will do the same, and post sentries in such manner as to ensure the security of the whole of the Troops during the halt. And the advanced Files and Flankers will occupy any heights or defiles that may be near.

THE HALT.

Flank Patroles.

68. When a Column is feeling its way in a country occupied by the Enemy, or is retreating in the presence of an Enemy, Flank Patroles are requisite to protect the Flanks

of the Column from sudden attacks or partial annoyance, particularly when marching through countries which are either woody, close, or hilly, having passes and defiles.

69. Flank Patroles to be of use in discovering the Enemy and to prevent a sudden Flank Attack (which on a march is the most dangerous of all attacks), must move at a sufficient distance to give the Column time to prepare for defence. It will, however, vary according to the nature of the country, whether open, intersected, or hilly. In most cases, 300 yards during the day, and about 150 in misty weather may be sufficient, so as it may be always able to observe the Column if no impracticable obstacles intervene. On meeting impassable impediments, such as swamps, &c., they will fall back near to the Flank of the Column, as by going round such objects they might risk being cut off.

70. Regard must also be paid to the description of the Enemy's force. For instance, if Cavalry is expected on a Flank when passing through an open country, or one thinly wooded, the Flank Patroles ought not to be at too great a distance and will keep a watchful look-out to observe them in time. They will on all occasions gain the crest of heights within their sphere of action, and examine all spots where an Enemy might be concealed.

71. The strength of a Flank Patrole will be in proportion to the length of the Column or the danger apprehended. In some cases one or more Companies may be required; in others, one subdivision may be sufficient, according as the Column may consist of a Battalion, a Brigade, or a Division, &c., or as local circumstances and particular occasions may render necessary.

STRENGTH.

72. If a Company is to form a Flank Patrole to a Battalion in Column of route, suppose to the right,—

FORMATION.

73. The Officer commanding the named Company will move it out to that Flank in double time. When at 200 paces, he will form subdivisions; the rear or left one will remain in Support, while the front or right subdivision will advance 100 yards, extending on the march at sufficient distances between files, so as to cover the whole Flank of the Column.

74. When the Column advances, the whole will face to the left, and move parallel with it, the leading File taking the outside Flank Skirmisher, or Flank Files, for the general

line of march, so as to have it always in view as much as possible; and the rear File ought to be within easy communication of the rear Guard. When the Column halts, the Patroles and their Supports face outwards towards the Enemy. A few Files may be thrown in advance at particular points as sentries.

75. In marching through defiles or a hilly country, some Files may be detached, diverging from the Flank, whilst others will crown the heights, forming a cordon, so that any suspicious place may be easily searched, or any concealed enemy be discovered.

76. On discovering the Enemy, either concealed or advancing on the Flank, the Flankers in this case must instantly fire, that the Commander of the Column may be acquainted in time of the danger. And the Officer in command of the Flank Patrole will immediately post it in the best position to meet and resist the Enemy, and make the most determined resistance, in order to give every possible protection to the Column while it is getting into a posture of defence. He will not retreat until recalled, or obliged to do so by an overwhelming force of the Enemy. If driven in, he will, in retiring, withdraw towards each Flank, to leave the front clear and enable he main body to fire immediately.

DISCOVERING ENEMY.

77. In general, only a Patrole on the Flank which is most exposed, or on which there is reason to apprehend an attack, will be necessary. However, on some occasions one on each Flank may be required, which was the case with the French army in Spain, where no detached Column could move with safety without them, to protect the Troops from sudden attacks of the Guerillas hovering in every direction. They were likewise necessary in the retreat of our army from Burgos. On several occasions, the French Cavalry were riding on both flanks of the Column, looking out for weak points or openings, always ready to dash through, in which they succeeded on one occasion, capturing the baggage of the Fourth Division.

78. These Patroles are exclusive of the Flank Parties, which on particular occasions may be necessary to crown heights during the passage of the Column through a defile.

Patroles.

79. The employment and judicious management of Patroles form an essential part of the precautionary duties of an Advanced Guard.

80. Patroles of Discovery are parties detached to such distances and in such directions as may be necessary, in order to feel for the Enemy; to obtain intelligence; ascertain his presence and positions; to reconnoitre the country with regard to roads, bridges, fords, defiles, &c., and report accordingly; also to keep up the communication between different Columns or Corps of the Army. They are likewise employed to examine farm-houses, copses, enclosures, &c., near the line of march capable of affording concealment to an Enemy, but too distant to be inspected by the Flanking parties.

PURPOSES.

81. They may consist of a Subaltern's party, or a Sergeant and twelve, or a Corporal and six men, according to circumstances. When sent for the purpose of gaining intelligence, small parties are preferable, as they have a better chance to approach near to the Enemy without being discovered. It is a rule that such Patrole never commits itself in action, or imprudently engages with the Enemy if it can be avoided, but ought to retire (under cover if possible) as soon as the requisite information is obtained.

COMPOSITION.

82. They must proceed with great caution and secrecy, ascending all heights from which a good view of the surrounding country can be obtained, and endeavour to discover everything without being seen themselves. The rule prescribed for the advanced Files and Flankers of the Van of an Advanced Guard are equally applicable in every particular to such Patroles, who will observe all those rules and precautions, as likewise the following:—

83. A Patrole on approaching a house, copse, or other object to be searched, will always endeavour to turn it by sending one or two Files on either side, out of range of musketry, which movement may induce the Enemy to leave it; whereas if the approach was made in front alone, it would be at the risk of losing men without an object. When the Flankers have passed round so as to command the Rear, a File of the Patrole will advance to examine it, followed at due distance by others to watch the motions of the preceding ones, and ready to give assistance, supported by the Reserve,

if required. When the advanced File is satisfied that there is no Enemy in the place, one of the men will make a signal by holding his firelock above his head in a horizontal position, when the others will move forward in proper order, and the Flank Files will resume their places.

84. On coming to a hill, the Patrole will halt; a file will be sent to each hand round the base, or, if extensive, at some distance on either shoulder. When this is done, another File will ascend to the top, taking care not to show itself on the summit, but will make observations from behind the brow, by lying down or creeping on, according to circumstances. If no Enemy is discovered, or in sight, a signal will be made as above directed, and the Patrole will move on.

85. The following is the usual disposition of a Patrole of Discovery on the line of march.

86. If the Patrole be of some strength, a double File will be thrown 100 paces in advance, with a File of communication half way, and a File about 100 paces to the right and left as Flankers, who will move parallel to the File of communication; likewise a double File to the rear, having a File of communication at 50 paces, or half way. But if the Patrole consists of only a small party, one advanced File 50 paces in front, with a File to the right and left, will be sufficient.

DISPOSITION.

87. Patroles may be employed to protect the Flanks of Troops engaged, especially when sent for the purpose of keeping a look-out, and cover the Flanks of an Advanced Guard. In such case they must be of sufficient force to offer resistance to the Enemy and keep him in check, should he attempt to turn or attack the Flanks.

Advanced Guard Pursuing an Enemy.

88. When the Enemy commences a decided retreat, either by voluntarily retiring from his positions, and endeavouring to steal a march upon his opponents, or in consequence of his having been defeated in battle, the Advanced Guard of the assailant or victorious army is destined to follow him up and continue the pursuit, in order to harass his rear, capture prisoners, baggage, &c., so as not to allow him, as far as due caution and prudence will permit, any respite.

89. In the first case, should it be suspected, from information received, or other circumstances, that the Enemy intends to retreat, his piquets must be narrowly watched. If it be

observed that larger fires than usual are kept up during the night, it may in general be considered as a proof that he is about to do so.

90. In order to ascertain how matters stand, an intelligent Non-commissioned Officer and a few trusty men without accoutrements, or anything except loaded muskets and bayonets, might be sent to creep forward with extreme caution, so as to get as near as possible to the Advanced Sentries or Videttes of the Enemy. The French often practice a stratagem of dressing up a figure of straw, with a chako, &c., and a pole for a musket, which, in imperfect light, have deceived people, so as to make it sometimes difficult to be detected until daylight. If this Patrole perceive figures of a dubious kind quite motionless, they should lie down, and watch attentively for several minutes; and if the figures remain stationary, two of the Patroles should creep on, listening to every sound. In this manner they may ascertain whether they are real or fictitious Sentries. Should they prove to be of the latter description, it will be found that the Enemy has decamped. Such an artifice was resorted to by the French in front of Santarem, when Marshal Massena commenced his retreat from Portugal. They have also, previous to retiring from Positions which they had in front of their opponents, erected sham batteries, as if they intended to make a serious attack, whilst at the moment, their baggage, sick, and Artillery, were perhaps in the act of retreating. An instance of this occurred the day after the battle of Busaco: they threw up during the night a large battery in front of our Position, but it was soon detected to be only a deception. Our telescopes told plainly that only an inconsiderable quantity of earth had been dug up, and so it proved, being a *ruse* preparatory to their evacuating their Position, which they executed the same night towards our left Flank. And nothing was visible the next morning but a few Squadrons of Cavalry, left to watch our movements.—*Lt. Col. Leach.*

91. When the Commander of an Advanced Corps has reason to suppose that the Enemy in his Front premeditates a Retreat, he will of course make all the necessary preparatory dispositions, and have all his Troops in hand ready to push on the instant that it is accurately ascertained that the Enemy has moved off, so as to overtake and follow up in the most vigorous manner the retreating foe.

92. In the second case, if the Enemy is defeated in action,

and driven from the field, and continues his retreat, the Commander-in-Chief, it may be supposed, will give all the necessary orders to the same effect. In either case the operations of the Advanced Guard for this particular service commences.

93. With regard to the operations of Advanced Guards in general, the principal ones which usually occur, or may be required in the field, are detailed in the several sections under the heads of "Attacking" or "Defending" defiles, bridges, woods, villages, &c., always bearing in mind the general rules.

94. That in making dispositions to attack an Enemy posted in such places or Positions, every endeavour must be made to turn his Flanks and threaten his Rear by flanking parties, previous to attacking him in front.

95. That no Advanced Guard or detached party on any particular service, after carrying a post, should ever Advance without first re-forming. Therefore the men should not be permitted to pursue the flying Enemy, but be rapidly re-formed, and wait orders to renew the onset in regular order.

96. Notwithstanding these wholesome regulations, Skirmishers are too apt to be urged on by daring gallantry, in the ardour of pursuit, or too apt to neglect all and each of them.

97. The following practical hints from Colonel Leach's excellent small work are deserving of the strictest attention.

98. In mentioning the duties of the Skirmishers of an Advanced Guard in pursuit of an Enemy retreating, when engaged in advancing through enclosures, woods, broken ground, defiles, or over heights, hills, &c., he says:—" A scattered line of Tirailleurs is usually attacked by advancing on them steadily but unceasingly, rooting them out from behind rocks, trees, hedges, or other cover, but especially by out-flanking them at the same time when practicable." Indeed dislodging the Tirailleurs of an Enemy from Positions where they appear to have most firmly established themselves *without an unnecessary loss of men*, may be reckoned amongst the many excellences which Light Troops ought to possess. Troops inexperienced in this desultory kind of service are apt, in Skirmishing, to move in little knots of four or five together, thus offering a decided mark for their opponents. So that, at the commencement of a campaign, many valuable men are thus lost to the service without necessity, before experience has taught the folly of grouping together, and the additional danger of not preserving their extended order, and also of not

making the most of such cover as the ground affords on which they are actually engaged.

99. When an Enemy retires in a hasty manner, or gives up ground which he does not deem requisite to maintain any longer, it frequently happens that the Skirmishers in pursuit are too prone, in the exultation and excitement of the moment, to keep up a rapid, and therefore too often an ill-directed, fire on the retreating party. This is not only an unnecessary expenditure of ammunition, but may prove disastrous to their own selves. How often has it happened that Light Troops in chasing the Skirmishers of their adversaries through woods, rough ground, &c., with much noise and smoke, and with a want of due caution, have run heedlessly on, and before they were aware what they were about, found themselves in front of a regular line of Infantry, posted behind some sort of cover, from which such a fire has been opened as to drive them back, and with considerable loss, on their own reserves. An instance of this kind occurred at the affair of Grijo, in front of Oporto, on the 11th of May, 1809; the French Skirmishers, after a sharp contest at the edge of the wood, rapidly retired, and leaped a stone fence some distance in the rear, closely followed by the gallant Light Company of the 29th Regiment. Those first up, in their haste, jumped over the wall right into a line of French Infantry lying there concealed, who instantly brutally bayonetted them. They had seized hold of an Officer (Stanhope), who had got on the top of the fence; but he was fortunately dragged back from their hands by his own men. The Enemy then fired a destructive volley; and on finding that their Flank had been turned, retired.

100. Skirmishers moving on in pursuit would act wisely (instead of rushing *pell mell* over the summit of each hill which the Light Troops of the Enemy have just abandoned) were they to make a temporary halt in order to rally and regain their order; while an Officer or Non-commissioned Officer with a few men should reconnoitre as expeditiously as possible the ground immediately beyond the brow, where it is probable they may find a well-formed body prepared to open a well-directed fire on the Skirmishers had they advanced; or some squadrons of Cavalry ready to ride in amongst them and to cut them to pieces in their loose extended order. Such accidents ought never to occur, and might be avoided if the Skirmishers on the extreme flanks, or the parties sent for the special purpose of protecting them,

kept a good look-out to secure their comrades from ambuscades or attacks in Flank, instead of joining with them to send random shots after the retreating Tirailleurs. It will be found that a few well-directed shots from the Skirmishers, when re-formed, are certain to cause the retiring party to accelerate their movements better than an irregular, noisy fire; which being ascertained to do little execution, will be comparatively unheeded. When Marshal Marmont, in 1811, introduced a convoy of provisions into Ciudad Rodrigo, Lord Lynedoch was posted near the river Azarva, with some divisions of Infantry and a small body of Cavalry, in order to check the Enemy in any attempt to penetrate towards the frontiers of Portugal. A large body of French Cavalry advanced towards this point, and became engaged with the British Dragoons, who, in conformity with orders, fell back skirmishing before a superior force. Lord Lynedoch placed some Light Infantry in ambuscade near the line of road by which his Cavalry were directed to retire; and so well was the affair arranged that the French Dragoons, following their opponents too confidently, and without a proper degree of caution, fell into the snare. A well-directed fire from the Troops in ambuscade arrested their progress, inflicting on them considerable loss, and threw them into such disorder, that the British Cavalry became the assailants, cutting down and making many prisoners in the pursuit.

101. In pursuing an Enemy in regions where the conflict is carried on principally between Light Infantry, it often requires some tact to dislodge him from certain rugged positions, where his Flanks are more or less secured by strong points, or from behind rocky mountain rivulets, especially if rains have swollen the fords, and if there are few bridges, and that the Enemy has taken advantage of such defensible ground, and is well covered by banks, rocks, trees, &c., overlooking the passages, so as to render it difficult to bring a fire direct on them from the front; and that it would require a considerable detour, occupying much time, to go round and reach their Flanks.

102. The urgency of the case will decide the Officer in command whether, at a certain sacrifice of life, the passage must be forced in face of the Enemy's marksmen, who, with their muskets, can fire with unerring aim; or whether it be more politic, by losing some time and fewer men, to dispossess him of his vantage ground by the adoption of other measures.

103. Should the banks be equally high, or more so than those occupied by the Enemy, and if it affords cover, no time should be lost in occupying it. In such cases, a large body of men should not be thrown forward at the same instant, which would expose them to a destructive fire in taking up their ground; but a few should steal forward at a time with rapidity, each man seizing on such cover as the bank affords, and selecting such spots as will best enable him to silence the fire of the enemy, and also protect those coming up to prolong the Line along the banks.

104. By this means the pursuing party may effect a lodgment near their adversaries, with but comparatively trifling loss. Indeed, well-practised Light Troops, by their knowledge of ground, and their tact in occupying it by one or two Files at a time, will in like cases suffer a third less than men unaccustomed to this description of warfare; and if any heights, rocks, &c., on one or both Flanks should happen to *trend* in such direction as would favour a Flank fire being opened on the Enemy, they should be occupied by Sharpshooters, which in most cases will be attended with the most favourable results.

105. Whilst a well-directed fire is thus kept up by the Skirmishers, a few discharges of grape-shot may be tried. And should there be a concentration of the Enemy's force at the only point by which a passage can be effected, some guns, if at hand, might be brought to bear upon them.

Skirmishers Advancing Pursuing the Enemy.

106. The following case in point of Attacking Troops in such positions may be given as an exemplification:—

107. During the retreat of the French Army, after the Battle of Vittoria, to Pampeluna, its Rear Guard, composed almost entirely of Infantry, took advantage of the various mountain streams to check our Troops in pursuit, who were consequently engaged in several warm Skirmishes. On one occasion the French Rear Guard had taken up a position behind a stream of inconsiderable size: its steep, abrupt, rocky banks swarmed with Light Troops. Heavy rains had rendered the fords impassable, and there appeared no other mode of dislodging them but by forcing a passage across a narrow bridge, or, at the expense of losing much time, turning one or both Flanks, should such be found practicable; but

which, by the delay, would have given the Enemy's main Column the advantage of gaining much ground.

108. Here the head of the British Advanced Guard (which only consisted of a few Companies of Riflemen and some Squadrons of Light Cavalry), which had followed the Pursuit, before the remainder of the main Advanced Guard, found the Enemy.

109. Count Alten, who commanded, observed that the French presented a considerable Force of Infantry in Line, in rear of their formidable chain of Skirmishers (and at some distance from the river) posted in such rough and tenable ground as to apparently bid defiance to all attempts at passing. He, however, instantly caused the bank on our side to be lined by Riflemen, with orders to keep down the fire of their adversaries as much as possible; whilst at the same time, he sent some Companies to occupy a rocky eminence a little higher up the stream, and situated on a bend in the river, sweeping well inwards towards the British side, which enabled our Riflemen, perched on the summit and sides, to take the Enemy in Flank, and to enfilade their opponents so effectually, from this height hanging so perpendicularly over the river, that they soon began to give way and quit their stronghold.

110. The main body of the Light Division was now seen hasting onwards to this scene of a truly Light Infantry conflict. The French Rear Guard was rapidly withdrawn, and the head of the British Van Guard was soon across the river, and once more in close pursuit of the fugitives.

111. It may be supposed that the fire of the Riflemen alone would not perhaps have sufficed to open a passage across the river, had not the appearance of reinforcements, supported by Cavalry and Artillery, tended materially to influence the Enemy in his retrograde movement; and that he would have held us, his pursuers, at bay much longer, had they not lost all their Artillery at the Battle of Vittoria. But it must be allowed that the flanking fire of the Rifles so effectually rooted out the French Sharp-shooters from their advantageous strong ground on the banks, that the primary obstacle was by this means removed.

112. It must be admitted that in the varied operations of Light Troops during a Campaign, all defiles, passes, rivers, &c., do not afford equal facilities to the Attacking party, as the above example; but it is sufficient to show that, when Light Infantry find themselves suddenly checked by obstacles

in Front, their tact and experience should be exercised to try what can be effected by operating on the Flank of their Enemy.

113. As a further example what well-disciplined and determined Light Troops may do, the following may be added:—At the action near Tarbes, in France, the 19th March, 1814, the French Rear Guard, consisting of about 5000 men, were posted in a strong position on a woody height, the lower ground in front being broken and covered with brush-wood. Our Advanced Guard, consisting of the 1st, 2nd, and 3rd battalions of the Rifle Brigade, were ordered to drive them from this position; for which purpose they all advanced in Skirmishing order, having the other brigade of the Light Division in reserve on their rear. The Enemy had covered their Front with a swarm of Light Troops, who occupied the hedges, fences, walls, &c., on the lower range of the high ground, which gave them a great advantage. But after much smart Skirmishing, they were all drawn in, and the Rifles gained the heights, where they found the French main body drawn up near a windmill on a steep acclivity, which allowed them to have two Lines, one immediately in rear of the other, so that the latter could fire clear over the heads of the Front one; nothing daunted by this formidable array, the Rifles boldly advanced to within 200 yards, each File seizing the best cover they could find, while their Supports lay down under banks, knolls, &c. They opened an animated fire, which was continued for nearly an hour, with great vivacity on both sides, when the Enemy at last were completely driven from one position to another, and then retreated with great rapidity to the plain in their rear. Thus did the Rifles alone, without requiring any assistance from the Reserve, succeed in defeating a large body of the Enemy formed in Close Order, and in a strong defensible position, with, comparatively speaking, no very serious loss (about 11 officers and 100 men), while that of the Enemy in killed, wounded, and prisoners, amounted nearly to the number of Rifles engaged, viz., 1200 men.

114. This would seem to prove the great advantage of rifles over the common musket, or else the superior mode of Riflemen using their arms beyond what is generally practised in Regiments of the Line. At all events, it demonstrates the utility of a ready and skilful taking advantage of ground and cover.*

* *See* Surtees' Twenty-five Years in the Rifle Brigade, p. 288.

Advanced Corps of Observation.*

115. When a Corps d'Armée, composed of all arms, is pushed on in front as an Advanced Corps of Observation to a considerable distance from the Main Army, for the purpose of observing the Enemy, who occupies in force an extensive line of country ; such, for example, as the Light Division was between the Coa and the Agueda on the frontiers of Spain and Portugal in the year 1810, under Lieutenant-General Sir Robert Crawford, which consisted of the 1st Hussars German Legion, 14th and 16th Light Dragoons, 43rd and 52nd Light Infantry, the Rifle Brigade, 1st Troop Horse Artillery.

116. In such a situation, all the merits of the Light Troops are speedily put to the test according to the degree of vigilance, zeal, activity and intelligence displayed by them in such an exciting and responsible service.

The position of the main body of this Advanced corps, whether chiefly held together in one camp or Cantonment, or whether otherwise disposed, and the general dispositions of the different sort of Troops employed, will depend on the nature of the line of country they are destined to watch, whether an open country with defensible places on one or both flanks,—whether a mountain range,—or whether a mixed position of posts, open, rugged, or hilly, with a river to cover the front.

117. In general it may be supposed that the main body will be placed in a central position, with one or more detachments to the right and left, to establish Outposts and support them, having free communication between each other, composed of Cavalry or Infantry, as the nature of the ground may require ; each post having in-lying Piquets always on duty, and in constant readiness to aid those in advance.

118. Should a river flow in front of the line, the number and nature of the different fords should be ascertained, and their depth practically proved. The most eligible points for the Enemy to construct Pontoon Bridges should also be observed and narrowly watched.

119. Piquets, either of Cavalry or Infantry, or of both, as may be best adapted to the ground, will be posted at all these fords, bridges, &c. The latter may be barricaded and crows' feet thrown into the fords to delay the Enemy ; but if the opposite bank is higher, so that the Enemy can command and sweep with Artillery such place and the adjacent ground

* From Lieutenant-Colonel Leach.

on our side, these obstructions so placed will prove of little avail. But Flag-staffs for day, and Beacons for night signals, should be erected on the nearest elevated spots, so that the alarm can be given along an extensive line from post to post, and to the rear in the shortest time. Any sudden bends of a river towards either flank of the line should be guarded with special care, because an Enemy, crossing suddenly at such points, by falling on the rear of Troops posted for the Defence of the convex Post, would seriously endanger their retreat.

120. If it be the Commander's intention to dispute the passage of the river, in the event of its being attempted by a force not much exceeding his own, he will probably place some pieces of Artillery, so as to command certain fords or bridges. Should there be any apprehensions of a night attack by a superior force, these can be withdrawn at nightfall, and moved up again at day-break. If, however, the Line of Defence is very extensive, having several bridges, &c., by which an Enemy might cross, and if these are only to be watched, and no serious opposition at every point intended, there may be no necessity of risking Artillery at the most advanced posts.

121. Should a fog come on, additional Sentries will be sent out, and the Patroles of communication kept in constant motion along the chain.

122. Advanced Sentries and Videttes ought invariably to be doubled, particularly in exposep situation, when in close contact with the Enemy, or at night; because a single one might be taken ill, or an evil-disposed one might desert to the Enemy. It also enables one to patrole with due caution occasionally to the Front, or towards the Sentinels on their right and left, each relieving the other at this duty, whilst one is always stationary, besides the confidence it inspires in the dead hour of night, when anxiously listening and looking towards the hostile posts.

123. The Piquets should have orders to examine and send to head quarters any suspicious persons endeavouring to pass their posts, either to or from the Enemy.

124. Cavalry Patroles of discovery ought, during the day, to be continually on the move to the Front and Flanks, and will prove a great security against the sudden approach of the Enemy; while the Cavalry advanced posts, if placed on knolls on ridges of hills, may, by preconcerted signals, com-

municate with rapidity any intelligence to the rear during the day. But an Officer ought always to transmit, with all speed, by a Dragoon, a written or verbal report of anything extraordinary.

125. As it is of the greatest importance to obtain every possible information of the Enemy's movements, an Officer should be stationed with a telescope on the most commanding height, top of the highest building or church-steeple which may overlook his principal Outposts or his Camp; to keep a constant look-out and eye upon him; and to make immediate reports in writing of anything going forward, to be transmitted by an orderly Dragoon stationed for that purpose. This was constantly done during the period the French lay in front of our Army, when in the lines of Torres Vedras. In certain positions during the daytime, more may be effected by all these means than by planting a multitude of Videttes and Sentinels, and who may thus be reserved fresh for night duty.

126. Although it is the paramount duty of Officers in command of Piquets to give notice of the most trifling movements of the Enemy, and the most timely intelligence of his advancing, still they must be careful not to send in hasty reports of Troops advancing without a certainty, as far as practicable, of their probable numbers. An inexperienced Officer not accustomed to Outpost duties, on observing a strong Patrole of Cavalry, accompanied by perhaps a small number of Infantry, may magnify it into something more formidable than it really is, and make a report accordingly. The consequences are that all the Troops are suddenly put under arms, the baggage hurried off to the rear, and much unnecessary trouble and inconvenience result from it to every individual of all ranks. These false alarms, if often repeated, have the ill effect of rendering the men tardy in getting under arms. And when a real cause of turning out arrives, they may perhaps not do it with the alacrity the case requires.

127. Still an Officer who makes an exaggerated report from sheer want of experience, and with the best intentions, errs on the right side.

128. It is the indispensable duty of a Corps of Observation to explore the country in which it is acting in every direction, and particularly to ascertain minutely every road, path, pass, or defile; and also the bridges or fords over streams in the rear in the line of country over which it would be obliged

to retreat, if driven back. Hence every Officer should always make himself thoroughly acquainted with all these particulars, so that he may know at once the road or path most eligible for his Piquet to fall back by, if forced, and the positions that will admit of keeping his pursuers in check, should this be necessary in order to gain time, and enable the Reserve to prepare.

129. As such an advanced Corps must be kept in the most effective state, ready for every emergency, all sick men, or those unfit for duty, should be sent off daily to hospitals fixed in the Rear, so that the Corps may never, in case of any sudden movement, be encumbered with the transport of sick. And as it is liable to be suddenly called under arms at any moment by day or night, every individual must be at all times ready to turn out on the shortest notice. Hence, exclusive of the Outposts and strong in-lying Piquets, one-half of each Corps not on duty, in exposed situations, should sleep alternate nights accoutred, and all on every occasion with their arms by their sides.

130. The Troops will always be under arms one hour before daybreak, and remain until after sunrise, or until it is manifest that no immediate movement is expected from the Enemy.

131. To ensure the assemblage of the Troops with alacrity, every one should know the spot he is to hasten to when the bugles and trumpets sound the "Alarm" and "Assembly." Hence every Company, especially if cantoned in villages, ought to have its own alarm post; and, as soon as formed, to be moved with all celerity to the Regimental one. And when all assembled, each Corps proceeds to the general rendezvous. And if in the immediate vicinity of the Enemy, a trusty man of each Corps may be sent at dusk to remain with the in-lying Piquet; so that in the event of an alarm, they may hasten to their respective Corps, to give notice and set the bugles sounding.

132. An advanced Corps being liable at all times to be suddenly attacked, such a contingency must always be considered as an occurrence extremely likely to take place. Therefore the baggage, weakly men, Commissariat, and all incumbrances must be prepared to move to the Rear on the first hostile moments being perceived.

133. As an example how an enterprising Commander on such service may surprise and defeat a skilful opponent,

the following dashing exploit may be recorded:—While the Duke of Wellington was engaged in the siege of Badajos, Sir Stapleton Cotton, now Lord Combermere, commanding the Cavalry, was then employed with it as an Advanced Corps of Observation in Spanish Estramadura to watch the motions of the Enemy, and to protect both the Covering and Besieging Armies. He conceived hopes of cutting off the French Advanced Guard of Cavalry opposed to him, under General Peyreymont, stationed between Villa Garcia and Usagre. On 10th April, 1812, having taken the necessary precaution, he ordered Anson's Brigade, under Colonel F. Ponsonby, to move during the night from Villa Franca upon Usagre; and General Le Marchant's Brigade from Los Santos upon Benveneda, to intercept the Retreat of the Enemy on Llerena. Ponsonby's Advance commenced the Action too soon. The French fell back before Le Marchant had gained sufficiently on their Flank to intercept them. They, however, drew up again in order of battle behind the junction of the Benveneda road. Le Marchant continued his march behind some heights skirting the Llerena road, which prevented the Enemy from observing him; Sir Stapleton, seizing this accidental advantage of ground, kept the Enemy's attention engaged in Front, by Skirmishing with Ponsonby's Squadrons, while Le Marchant, secretly pressing on at the back of the heights (unobserved by the Enemy, who had thrown out no Flank parties or Patroles of Discovery), completely gained their Flank; and when Ponsonby charged them in Front, he sent the 5th Dragoon Guards to attack them in Flank. The Enemy were entirely overthrown, and fled in confusion.*

134. When an Advanced Corps has to retreat, whether it be ordered to do so or whether obliged to retrograde in consequence of the Enemy pushing on with superior numbers, in either case the execution of the movement will be conducted in the same manner as above explained for Rear Guards.

135. In some positions there may be several roads which lead to the same point in the line of Retreat. The Commander will in general give previous orders, if any of the Piquets, or any portion of the Troops, are to retire by any of such roads, to re-unite the Column at a fixed point, or whether the whole shall move in mass.

* See Napier, vol. iv., book vi., p. 438.

SECTION XIX.

Rear Guard.

1. A REAR GUARD is a body of troops which, when an Army or other body retreats, marches the last, so as always to be interposed between the Enemy and the main body. The following comprise its principal duties:—

GENERAL PURPOSES.
2. To observe the Enemy, and cover the retreating Column, to prevent him from pressing too closely, or from stealing round, and gaining on the Flanks of the main body.

3. To keep the Enemy in check, by obstructing his advance, and avert any sudden assault in the Rear, so as to secure the Retreat from interruption, and give the Column time to gain ground, and surmount any obstacles or difficulties it may meet with on the route.

4. To hinder the Enemy from capturing baggage, stragglers, or any part of the *matériel* of the Army.

5. The object, however, of a Rear Guard is not so much to engage the Enemy as to keep him at a distance, and guard the Retreat from annoyance. Resistance will generally cease when that end is attained. However, when the pursuit is close and vigorous, every hedge-row, copse, ridge, bridge, or defile, &c., becomes a post which must be defended with obstinacy.

6. The strength of a Rear Guard depends upon, and can only be determined by, the force of the Enemy, the nature of the country, and the degree of resistance that may probably be required to give the Column time to proceed without interruption. When required on a large scale, a Corps d'Armée similar to an Advanced Guard is usually appointed for this duty.

7. But its composition, whether it is to consist of Infantry or Cavalry, or both with Light Artillery, will depend also upon

the composition of the force employed by the enemy, and the nature of the country. In an open country and extensive plains, the Rear Guard will generally be composed of Cavalry supported by Infantry. In an intersected or hilly country, the Infantry will constitute the principal force, and in either case they will be supported by Artillery.

8. In a mountainous or woody country, or those intersected with canals, dykes, &c., the Cavalry will be sent to the Rear, and the Infantry alone will perform all the duties, having only a detachment of Light Horse for the purposes of patroling when required; remaining in the Rear to observe the Enemy, and give notice of his near approach, or to carry despatches and keep up the communication with the main Column.

9. The disposition of a Rear Guard is exactly that of an Advanced Guard reversed. Therefore, when any body, whether large or small, is ordered for this duty, it will be previously countermarched so as to stand faced towards the Rear, and ready to oppose the Enemy.

10. Rear Guards may be considered under two heads, according as the Column may be advancing or retreating, and the formation will depend:

First,—Whether it is the usual Rear Guard of a body advancing, to be formed simply as a Rear Guard on the line of march following a Column " en route," as directed in Her Majesty's Regulations, Part V., Sect. v., No. 3.

Second,—Or whether it is a Rear Guard destined to cover an Army in retreat, and to be formed in fighting order prepared for defence. Under this head the immediate preparations for retreat will depend whether the Army whose retrograde movement it is to protect commences its retreat voluntarily, whether occupying a country or any particular position; as, for example, Sir John Moore's Retreat in 1809 to Corunna, Massena's Retreat from the Lines of Lisbon in 1811, or Lord Wellington's Retreat from Burgos in 1812; or whether the Army, having been defeated in the field, is obliged to retire in the immediate presence of the enemy; such, for instance, as the French were at the Battles of Vittoria and Waterloo.

11. In the first case, this operation, if unsuspected by the adversary, may probably be effected without much difficulty or danger; and if the intended movement is kept profoundly secret until the moment of executing the order, the chances are that the Commander may be enabled to steal a march upon the Enemy.

PRECAUTIONS BEFORE RETIRING.

12. The following hints from Colonel Leache's excellent little work, "Recollections relative to the Duties of Light Troops" are very applicable.

13. The inhabitants may be permitted to pass and repass the Lines, as usual, until immediately before the Retreat is about to commence; because, if all communication was cut off for one or two days previous to the movement taking place, the suspicions of the Enemy would be raised. But to ensure perfect secrecy, directions may be sent to the piquets some time before sunset to prevent any of the inhabitants from passing over towards the Enemy's lines, but to permit free ingress from thence to our own. The Troops may afterwards be warned to march; and the baggage, commissariat, and weakly men, if any, sent to the rear at nightfall, followed by such portion of the Artillery as may not be deemed necessary; and the retreat of the main body commenced as soon after dark as may be thought expedient. The nature of the country will determine the order of the different sorts of Troops in retiring, and the dispositions necessary to be made.

14. If very open, or in large plains, the Cavalry would protect the movement and cover the Retreat. If close, intersected, or hilly, &c., the Infantry would perform that duty.

15. The Piquets will remain holding their ground, and orders will be sent to them not to retire until a certain hour, which will be named. When the time indicated arrives, they will be careful to retire with the greatest punctuality, recollecting to replenish their fires, by piling more wood on them previous to their evacuating their posts. They may likewise leave figures of straw posted as Sentries, or other devices usual on such occasions, as a *ruse* to deceive the enemy. The piquets well then move off in profound silence, and should there be any turf or soft ground in their line of march, they will (especially if they are Cavalry) move on it in preference to a road, where the noise of horses' feet might be heard by the Enemy's Piquets. On getting clear off, they will accelerate their pace to overtake the main body, a portion of which may have formed up at some distance in the Rear, to protect them in case of their having been attacked.

16. Should these precautions be attended with success, and the Retreat not have been discovered by the Enemy until the following morning, it may be concluded that the Column will have gained sufficient distance to preclude much apprehension of its being overtaken, perhaps for a day or two, or, at

all events, until it has secured good defensible positions, in order, if necessary, to resist and check the pursuing Foe.

17. Should there be several roads leading to different places on the general line of Retreat, it is often good policy, when circumstances admit, to make a portion of Troops retire by one or more of such roads, exclusive of that taken by the main body, with orders to join it, if such should be necessary, at certain places indicated. The object of this is to leave the Enemy in uncertainty of the true line of Retreat; and as he must lose time in endeavouring to gain information, or in sending out Patroles to ascertain the true direction, it will be so much time gained to the main body. No stronger instance of this can be given than that of Blucher's Retreat on Wavre after the Battle of Ligny, 16th June, 1815; on which occasion Marshal Grouchy, who was sent in pursuit of him, lost, even according to the French accounts, nearly a whole day in finding out whether the Prussians had retreated towards Liege, Namur, or Wavre. This proved of the most vital importance. It gave the Prussians time to retire unmolested, to re-form in good order, and to form a junction with their Allies, and next day to retrieve their recent defeat by an immortal victory.

18. But although a Rear Guard may thus, unmolested, evade the Enemy, it can only in the long-run expect but a temporary respite: an enterprising Enemy will press rapidly on to overtake it. The Commander ought therefore to be at all times prepared to receive him by a proper disposition of his force in fighting order, and will march with the same precautions as if the Enemy were actually close upon his Rear.

Covering Retreat.

When the Enemy at length comes up, the arduous and important duties of a Rear Guard really commence, as hereafter given, viz.:—

19. Under the second case,—When an Army has been defeated in action, and is obliged to retreat in the immediate presence of the Enemy, under protection of a Rear Guard; or when an advanced Corps of Observation occupying a line of Posts is attacked at all points by a superior force, and is under the necessity of retiring fighting before overwhelming numbers.

20. The preparations to cover a Retreat in immediate presence of the Enemy should, on all occasions DISPOSITIONS. when practicable, be made in the Rear. The Corps d'Armée destined for this service (usually a portion of

the Reserve) whether composed of Infantry or Cavalry, or both, with Artillery, will, in general, be posted in a favourable position previous to the main body's retiring; so that under its protection the Columns can be withdrawn, and commence their retreat in due order. With regard to the disposition of the Rear Guard, whether the Cavalry alone should cover the Retreat, or whether this duty should be performed by the Infantry, or by a portion of both arms acting in conjunction with Horse Artillery, must depend on the nature of the ground, or that of the sort of Troops employed by the Enemy; while the immediate dispositions to receive and resist the pursuing Enemy will generally depend upon those he may make to attack, whether in Close or Skirmishing order. If in the latter, such proportions of Light Troops (whether Cavalry or Infantry) as may be deemed requisite, may at once be thrown out in Skirmishing order. Or if that is not required at the moment, they may be so disposed as to be in readiness to do so at any time on the march as occasion may offer.

21. A Rear Guard covering an Army is supposed to have the direction of its retreat previously cleared by the Army or Column which precedes it. Hence all the principal attentions of the Commander is directed towards the Enemy in his rear. But his foresight will induce him (particularly during operations in an Enemy's country, where the population is hostile and acts in armed bands) to send a detachment of such magnitude as the case may require, to keep up the communication with the Army or grand Column; to reconnoitre and send back Reports of the nature of all defiles, passages of rivers, &c., in the direction of the march, and to occupy such places if necessary to cover his approach; likewise to repair such bridges, roads, &c., as may have been damaged by the Army or the peasantry.

22. Should circumstances occasion a long interval to occur between the Rear Guard and the Retreating Army, such detachment should be of sufficient strength to open its way by main force, or at least be able to resist and prevent, if possible, any sudden attack, should armed bands of the Enemy, by a detour, have succeeded in getting between the Army and its Rear Guard. Instances of this occurred to the French Army in Spain by the Guerillas; and such was the case at Waltersdorff in 1807, when the Prussian Corps under Bulow was cut off from the Prussian Army by the Corps under Marshal Ney, and had to open a passage by force of arms.

23. The great object of a Commander's attention is to observe the movements and near approach of the hostile force, so as to be able to choose the most advantageous positions to arrest his progress if necessary. Should the Enemy press too rapidly on, he must offer the most determined resistance, whatever the force of the Enemy may be; or should he find himself obliged, in order to gain time, to resist the Enemy, he may take an opportunity of doing so after passing through a defile, bridges which cannot be turned, or a village in certain localities. For this purpose he may form up or deploy a due portion of his force to receive the Enemy, and attack him as he debouches, because it must be recollected that Troops deployed in front of a defile to resist an Enemy debouching possess the advantage of presenting a greater extent of front than an Enemy, who is of course obliged to regulate his by the breadth of the defile, bridge, &c. Hence, should the Enemy venture to pass through, and push rapidly on, he is liable, as soon as the head of his Column presents itself in coming out, to be taken in Front and Flank, and enveloped before they can form, and driven back in confusion.

24. The following example seems very appropriate:—In May, 1811, Sir William Lumley occupied Usagre, in Spanish Estramadura, as an Advanced Post with a Brigade of Cavalry. This town was situated on a hill, and covered on its rear towards Los Santos (our side) by a river, with steep and rugged banks, and had only one outlet, viz., by a bridge over this river. Marshal Soult had directed Latour Maubourg, with a large force of Cavalry, to seize upon Usagre, and to scour the country beyond it. On his approach, Lumley retired across the river; the Enemy soon afterwards occupied the town, and their Advanced Guard proceeded to follow him up. A body of French Light Cavalry moved along the right bank with the intention of crossing by a ford lower down, in order to cover the passage of the heavy horsemen. But before they could effect this object, General Bron rashly passed the bridge with two Regiments of Dragoons, and drew up in line just beyond the bridge, without having taken sufficient distance in advance to allow room for the Reserve to debouche and form up. Lumley was, during this, lying close behind a rising ground in their front. Immediately on the French commencing to advance, our Artillery, rapidly moving up, opened a destructive fire upon them. The 3rd Dragoon Guards and 4th Dragoons vigorously charged them in Front, while General Maddere's

Portuguese Cavalry at the same time attacked them in Flank. The Enemy was overthrown at the first onset, and fled towards the bridge, which, being choked with the main Column of the French Cavalry, the fugitives were impeded in their retreat; they turned to the right and left, and endeavoured to save themselves amongst gardens on the banks of the river, where they were pursued and sabred, until the French on the opposite side, seeing their distress, dismounted a portion of Cavalry, lined the banks, and opened a fire of carbines and artillery across, which checked the victors."*

It may be also necessary on some occasions to defend the gorge of a defile in order to keep an Enemy in check, and afford the main Column sufficient time to effect the passage. During Massena's retreat from Portugal, the French army, which retired in one massive column, was long delayed in an intricate defile in rear of Redinha. The gallant Ney, who commanded their Rear Guard, with his wonted tact and judgment, on the 12th March, 1811, took up a strong position on some heights in front of the village and bridge of Redinha, where the defile began; and although closely pressed by our Advanced Guard, he maintained himself with great resolution until the desired object was attained, viz., the safe passage of the Column. He then withdrew his force with surprising celerity and in admirable order.†

25. When a mixed Rear Guard retreats through an open country, the greatest part of the Cavalry may be formed in the rear, so as to give the Infantry a portion of its time to gain a height, wood, or other locality, and then the Cavalry can retire under cover of the Infantry so posted. In the same manner, when a bridge or short defile, &c., is to be passed, the Cavalry may deploy in the plain, in front of such obstacle until the Infantry has passed; and then the Cavalry will follow, and pass with the greatest celerity, under cover, if necessary, of a detachment of Infantry left for that purpose.

26. When a village occurs in the line of Retreat, the immediate possession of which, from its local situation, might afford the Enemy an advantageous opportunity of annoying the Rear Guard, it may be manned with Light Infantry; and the Troops will pass round it, by one or either side if practicable; but if not, they will proceed through it as quickly as possible, so as in no case to mask the fire of the Light Troops

* See Napier, vol. iii., book xii., p. 554.
† Ibid., p. 464.

posted on the commanding points or houses affording, from their position, a flank or cross fire. The Cavalry, or such portion as may be required, will form in the rear of it, to protect the retreat of the Infantry.

27. Although, during a retreat, the greatest portion of Light Artillery attached to a Rear Guard would generally, if traversing a strong country, be sent on before, still, when defiles have to be passed, affording favourable position (whose flanks are perfectly secure, and which can neither be attacked nor turned without the Enemy making a long detour) and where it is deemed necessary to hold the Enemy in check. Decided advantage may be obtained from this arm on such occasions, where such portions of it, as circumstances dictated, might be placed to defend the retiring Troops, particularly if the guns are supported by a body of Riflemen to co-operate with them in the defence of such passes, defiles, fords, bridges, &c.; and not only this, but the passage of such particular obstacles might be rendered very difficult for the pursuers, so as to retard their progress, and much time might thus be gained. As soon as the Enemy showed a determination to carry such position by main force, at all hazards, regardless of the sacrifice of life, the guns will be withdrawn under cover of the Sharp-shooters, and the retreat continued. The further order of march, however, will entirely depend upon and be regulated by the features of the country to be traversed. Indeed, there are instances wherein certain positions in a strong or mountainous country, such as the Tyrol, which could only be approached by one road, that a gun might be so placed as to pour down grape-shot with deadly accuracy on an invading foe or pursuing enemy, dealing death and destruction until the hostile column has been torn and shattered by its fire, and that of Riflemen, before it gave up the point.

28. A Rear Guard is a service of the utmost importance, the duties imposed upon it being of the most arduous nature; so much so that, in fact, on their being well or ill executed the fate of a whole army may depend. It is only by uniting great intelligence with intrepidity and firmness that a Commander can hope to succeed in conducting it with credit to himself and uphold the honour of the Army. The Regiments selected for this service should be those who bear the highest reputation for order and discipline, and on whose good conduct every confidence can be placed under the most trying circumstances, when the energies and activity of every individual are required to endure the privations, danger, and

often lengthened or rapid marches incident to such operations. "The soldiers should be reminded that it is during a retreat that the discipline, spirit, and good order of Regiments are incontestibly proved.

29. "It is in keeping well together, in obeying implicitly the order of their Officers, in not straggling from their ranks, nor committing excesses, by plundering or in getting intoxicated, that their excellence is put to the test, that it is these qualities which render them formidable, under the most trying circumstances to an exulting and pursuing Enemy."

30. Such are the general principles on which the operation and duties of a Rear Guard are usually conducted; but without entering into further details of them here, or of those of the Cavalry and Light Artillery, which will be found under the Sections.

31. The following is principally confined to the general duties of Light Infantry and Skirmishers on such service :—

32. When a body of Light Troops is destined to cover the retreat of a Rear Guard, or other body, whether in Columns or Lines, in presence of the Enemy, the preparations for this movement are usually made in the Rear, so as to be concealed from his view. Therefore when Light Troops are thrown into Skirmishing order for this purpose, it may be done as follows:—

33. Suppose a Battalion is to be employed, it will, previously to the Troops falling back, take post in a proper position in their rear, and may there extend two, three, or more Companies as Skirmishers, leaving at due distance an equal number of Companies in support; and the remaining ones will proceed on to form the Reserve in rear of all, having likewise thrown a Subdivision out on each Flank of the Chain to keep a look-out and protect their Flanks. When the Column or Lines have passed to the rear, the Light Troops so posted become the Rear Guard, and will follow when the Troops have gained a sufficient distance.

34. In some instances a retrograde movement may be masked under a demonstration of attack; in which case the Chain of Skirmishers may be thrown out in front of the Columns or Lines, while the Supports and Reserve proceed to take post at their due distances in rear of the Lines. The Troops may then be rapidly withdrawn, and the Rear Guard will remain formed, ready to act as circumstances may require. Should the Enemy push on, the Skirmishers will open their fire, and will commence retreating, when it is judged the main

body of the Rear Guard may have got sufficiently on the road. If he does not immediately come on, they will likewise retire at the due time, but without firing, at the same time keeping a sharp look-out occasionally to the rear.

35. The Flanks of a Rear Guard ought at all times to be well secured by Flank parties or Patroles, in order to guard against being out-flanked, or the Skirmishers cut off, particularly in passing through woods, a very intersected country, defiles, &c. While a Rear Guard is disputing any point, the Flank parties must be on the alert to notice any attempt of the Enemy to turn its Flank, as, for instance, by fords, either above or below, when a stand is made at a bridge; because it may be presumed that an Enemy will always avail himself of every opportunity of stealing round the Flanks of his opponent.

36. An Officer in command of a Rear Guard will be cautious not to engage unless when forced to do so; but will hold every strong position or defile as long as practicable without committing himself, so as to enable the Column to gain ground. He will at the same time not suffer himself to be too far separated from it. On leaving a position, he must gain the next strong ground as expeditiously as possible, so that the men may never remain exposed in open places.

37. He will endeavour to avoid being hurried into a defile in presence of a superior force. To obviate this, he ought to send an intelligent Officer to the rear to fix upon advantageous positions for making a stand, where the Supports, or a sufficient portion of the Reserve, may be posted in Skirmishing order previous to the arrival of the old Line, which, after passing through the new Chain, will, when at due distance, either form up as Supports, or proceed to join the Reserve, as the case may be. He will thus continue the retreat from one position to another by alternate Lines, relieving and supporting one another while retiring.

38. On leaving a defile, or very intersected country, and coming to a plain, should there be no danger from Cavalry, the Skirmishers will hold the last position or enclosure until the Reserve and Supports have retired into the plain, beyond the range of musketry, when the Supports will extend and lie down. The old Skirmishers will then give up the position, running rapidly to the rear, and then form up as Supports to the new Line.

39. But should there be any apprehension of Cavalry, the

Reserve, on leaving the defile, bridge, &c., will halt in front of it to draw off the Skirmishers, while the Supports will retire beyond the Reserves, and will assemble in Column of Sub-divisions. The Skirmishers will be recalled by a given signal, and will proceed to unite with the Column of Supports, which will continue retrearing. The Reserve will then fall back and follow. The whole will thus, in two separate bodies, mutually supporting one another, and always ready to form Squares, continue the retreat in safety across the open country. On reaching strong ground or enclosures, the former Supports will extend to line the fences or other position, and the original formation of Skirmishing order will be again resumed.

40. There often occurs, during a retreat through a hilly or mountanous country, favourable defensive positions (such as certain passes which cannot be turned, or rivers having commanding rugged banks on the retreating side) where Riflemen may be employed with advantage, particularly when acting in conjunction with Light Artillery in protecting Troops, and for which they are especially suited. On such occasions they must occupy every inch of ground from whence their fire can be delivered with effect. Their weapons will, under all circumstances, cause more destruction than any other description of small arms, but it is particularly when posted for the defence of certain points of a position from whence they can take aim coolly with a steady hand, resting their rifles on a rock, bank, tree, &c., that their fire is so deadly. Well-trained Riflemen thus posted ought to expend few shots without effect. During a long-continued skirmish in woods, or amongst rocks, enclosures, &c., when heat, fatigue, and excitement tend to render the hand unsteady, the fire of Riflemen or other Troops can never be so accurate as when stationary and firing from a rest.

41. All Skirmishers when covering a Retreat ought to reserve their fire as much as practicable, and instead of a wild straggling tiraillade, opened as it were merely to return shot for shot with their pursuers, that they, when favourable opportunities offered, should wait patiently with rifle or musket rested against a tree, bank, &c., and to make every shot with effect; they would thus more decidedly check the ardour of their pursuers, and make them cautious of pressing on too closely, than by numbers of cartridges, consumed in a hurry and in a random fire.

42. The effectual protecting the retreat of their army with-

out unnecessary loss of men may be esteemed one of the highest qualifications of Light Troops.

43. In retiring, the Skirmishers ought frequently to look behind them, and to either Flank, that the Enemy may not fall upon them unawares. They will at all times be prepared to form Squares in case of a rush of Cavalry, or to leap over the fences on each side of a road so as to take Horsemen in Flank should they dash on. The Officers will be careful to prevent any Skirmishers or men of Flank parties from falling into the rear or loitering in any houses or villages.

44. When a Rear Guard halts, the whole must turn to their proper Front, facing towards the Enemy, so as never on any occasion, when halted, to remain with their backs towards him.

45. On all occasions during a Retreat, the Officer in charge of the Reserve of a body in Skirmishing order, on coming to a river, swamp, or other obstacle in the Rear, ought to send notice back that the Supports and Skirmishers may be conducted in retiring in the proper direction to such fords, bridges, paths, &c., as are most suitable to pass these obstacles in the most convenient manner.

46. In like manner, should the column or main body of a Rear Guard, &c., meet with any obstacles, such as a narrow pass, a difficult ford, or broken bridge, &c., which would impede for a time the rapidity of its progress, notice ought to be sent back to the Rear Guard in order that it may hold the Enemy in check during the delay which must ensue. Should the Enemy press on vigorously, the column may leave a detachment at such obstacle, defile, &c., to prepare the defence, and to protect the Rear Guard in passing over it, or to assist, if necessary, in maintaining the post as long as may be expedient, in order to afford relief to the main Column and enable it to gain ground.

47. Every Rear Guard should be provided with entrenching tools to enable him to barricade bridges, or to break up roads, &c., so as to throw every impediment in the way of a pursuing Enemy.

Rear Guard in Line of March.

48. With regard to a simple Rear Guard in the line of march following a body advancing in column of route, it is formed as directed in Her Majesty's Regulations, Part V., Sec. v.,

No. 3, for an Advanced Guard of a single company only in reversed order.

49. The Company or other body ordered will be countermarched, and remain halted facing to the Rear, until the Column, Baggage, &c. have gained a sufficient distance, when the formation may take place in succession to the Rear, from Right to Left Flank, according as the Column is marching Right or Left in Front.

50. In this case, a Rear Guard on the line of march will act as a Police Guard to collect stragglers, to prevent plundering, to see that no baggage remains behind or is lost; and when advancing in an Enemy's country, it ought to be of some strength in case of marauding parties of the Enemy or armed peasantry endeavouring to cut off the baggage or stores, in which case, two or more Companies may be employed; Companies or Subdivisions, as the case may be, acting as directed for Sections.

SECTION XX.

Attacking Bridges.

1. In the attack of bridges or other small defiles, Light Troops forming an Advanced Guard, or covering the advance of a Column, may be employed with great advantage. As the Skirmishers open and prepare the way for the seizure of the bridge, &c., the execution of the movements will, however, depend whether the Enemy disputes the passage or not, and other circumstances.

2. Should the Enemy be in force, and defend the bridge, it must be carried by force under cover of the Skirmishers. For this purpose the Chain will advance close up to the bank of the river, to right and left of the bridge, when they will get under cover, and continue to keep up a brisk fire upon the Enemy. Should the ground be so exposed that they could not advance without serious loss, and still less retain their position, the men may take off their knapsacks and carry them before them in their left hands, dash rapidly up, and lie down, placing their knapsacks before their heads, and open their fire.

3. When the strength of the Enemy and the nature of the defence have been ascertained, such as whether barricades have been erected or planks of a wooden bridge have been taken up, or that no serious obstacles occur, and that a passage is practicable, the Supports having been closed in and formed in Column, will proceed to carry the bridge by storm, and force the passage with the bayonet, supported by the Reserve, which, in case the first attack should fail, will immediately send forward a portion to renew it. When the bridge has been carried, and the Reserve crossed over, the Supports will extend in such direction as may be required

to pursue the Enemy, according as he may retire, either direct, or turn off to the right or left. The Reserve will continue to maintain possession of the bridge, until the old Skirmishers, having assembled and crossed over in double time, have proceeded to form Supports to the new Line: the whole then move forward, according to the original formation. But if the Enemy do not give ground readily, the old Skirmishers will not pass over, but remain in their position on the banks, until sufficient ground has been gained; so that in the event of the Troops being driven back, they might have time to resume their former position on the banks, and open their fire. In carrying a bridge, a venture may be made to gain as much ground as possible with all expedition. Therefore, when practicable, a large force of Skirmishers should be displayed under cover of Cavalry, so as to be able boldly to remain in advance, or to push on.

4. And on all such occasions, parties will previously be sent to either hand, above and below the bridge, to pass over when practicable, by fording, boats, &c., in order to turn the Enemy's Flanks. And even if there is no ready means of passing, still they ought to make a demonstration of doing so.

5. On approaching a bridge, should the Enemy have Troops remaining on this side of the water, and if these are to be attacked in withdrawing, a bold and decided dash should be made, with such description of Troops as may be necessary, according as the force of the Enemy may consist of Cavalry or Infantry.

6. If the bridge is but feebly occupied, or being only observed, the Skirmishers may take post on the banks, while the Supports, in close order, dash over at once, followed by the Reserve, and effect the passage. If no resistance is offered in advancing, the Skirmishers may gradually close, face inwards, and at once dash over, extending again after crossing, followed by their Supports and the Reserve.

Defending Bridges.

7. When Light Troops are employed during a Retreat in covering the passage of bridges, defiles, &c., or in defending them, the following disposition may be observed:—

8. On approaching the bridge, the Skirmishers hold the last position until the Reserve passes over, and has taken post at the bridge-head, or the out-let of a defile; when it will immediately throw out a new Chain of Skirmishers to line the river to the right and left; while the Supports will close in, and form in Column in front of the bridge, until the line of Skirmishers are withdrawn, which is effected by the Skirmishers inclining inwards while retiring, and, when near the bridge, rapidly withdrawing by both Flanks of the column of Supports. They will, after crossing, form up in rear of the Reserve. The Supports will then follow, and in like manner join the Reserve.

9. The new Line of Skirmishers will open their fire as soon as their front is cleared, and parties will be detached from the Reserve to both Flanks to watch the fords or such places where the Enemy might attempt to effect a passage.

10. The whole are thus prepared to defend the bridge, or to retire, as may be ordered. If the Retreat is to be continued, Supports will be thrown out from the Reserve, and the whole will retire in proper Skirmishing order.

11. Formerly the Skirmishers used to be the last to retire across; but experience has proved that they ought not to assemble or make a stand in immediate front of a bridge. They can make no effectual resistance, and should the Enemy overpower them, he might pass *pell mell* with them, while they would prevent the Troops posted on the other side from resisting him.

12. This is the general mode of effecting the passage of a bridge on common occasions.

13. But during a retreat, should the Enemy press on too rashly, a stand must be made in front of the bridge. For this purpose, the Skirmishers should be reinforced and supported by Cavalry, so as to keep him in check, or even for a time to assume the offensive, in order to afford time to the retiring Column to defile over, and to make the necessary dispositions either to barricade or destroy the bridge, which, if built of stone, may be barricaded by waggons, carts, dung, &c. Or if such materials are not at hand, the pavement can be taken up, and a breastwork erected with it. The removing of these obstacles under fire will cost the Enemy much time, and a great loss of men.

14. If the bridge is of wood, it can be more readily rendered impassable, because a part or the whole of the planks may be taken up, and either thrown into the water or piled up as a barricade to protect the Skirmishers, who can easily replace them, should they drive off and have to follow up the Enemy. Or if the bridge is to be burned, the planks may be placed in loose heaps at intervals; and then these, with the rails and frame smeared with tallow or pitch, or sprinkled with oil, and the whole covered with straw or brushwood, and then set on fire. In dry weather, gunpowder strewed about, and a train laid, may be employed for rapid conflagration. On such occasions it may happen that the Skirmishers will have to remain to the last moment in covering such preparations. Therefore, boats ought previously to be secured to bring them off, and the Cavalry will have to pass by swimming.

15. If the defence of the bridge is to be continued, the main body of the Infantry Reserve will be posted at about 150 paces distance, leaving parties at the head of the bridge to repulse any small bodies of the Enemy who may attempt to remove the barricade, or venture to pass over. But should a large force of the Enemy succeed in forcing a passage, these detached parties will withdraw to either hand ready to take him in Flank. And as soon as a certain portion of the Enemy have cleared the bridge, the Infantry will open their fire by giving a volley or two, and then charge with the bayonet, so as to drive him back before he has had time to form. The Skirmishers who were posted along the banks will, immediately on the Enemy's attempting to cross, direct their fire against the attacking Column. Should there be any mills, houses, rocky heights, &c., in the vicinity, they will of course be occupied, and every advantage taken of them to promote and prolong the defence, according to circumstances and the judgment of the Officer in command; while the Flank parties will use the utmost vigilance to observe the Enemy, and prevent his attempting to pass the river, either above or below the bridge. By these means, particularly if supported by Artillery, a bridge may be defended for some hours.

16. For example, a Rear Guard consisting of one Division, composed, suppose of one Light Corps, four Battalions of Infantry, four or more guns, and two or more Squadrons of Cavalry, retiring before the Enemy, may effect a retrograde

movement across a river in the following manner:—One Battalion of Infantry, one Squadron, two guns, and a wing of the Light Corps will be left to cover the passage. This detachment will take post on the most advantageous position in front of the bridge, and will endeavour to keep the Enemy in check until the main body has retired across, and taken up a position on the opposite side in the following order:—

17. Two Battalions of Infantry will be posted about 150 paces beyond the bridge, having their outward Flanks obliquely thrown forward, and the other Battalion in Column of Reserve in their Rear. Two guns will be placed in the interval between these Battalions, and immediately in front of the bridge, so as to command and rake it, and the Cavalry in Rear of both the outward Flanks of the Infantry, while the wing of the Light Corps will send out sufficient Skirmishers to line the banks on either hand of the bridge, but not too close to it, having two Supports near the head of the bridge. Part of these may occupy any mills or houses, if such are favourable for defence.

On these arrangements being completed, the Artillery left on the other side will retire across, followed by the Cavalry. The Battalion of Infantry will take post in Column in front of the bridge, under cover of which the remaining wing of the Light Corps and Skirmishers will rapidly be withdrawn, when the Column will also retire. All these Troops as they pass over in succession will, immediately on clearing the bridge, move with celerity to the right or left as the case may be, in order to leave the front of the Troops posted in position open. Thus the wing of the Light Corps, with the old Skirmishers, the two guns and Battalion of Infantry which retired last, will proceed to the Rear and form up with the Reserve; while the Cavalry will join the others in Rear of the Flanks of the Infantry.

18. Should the Enemy only push on in small parties, the Supports near the head of the bridge will be ready to drive them back. If he press on in force, the Supports will withdraw to the right and left, and the Skirmishers nearest the bridge will close to their outward Flanks (so as to admit of the guns and Troops opening their fire), while they will fire obliquely to take him in Flank. When the Enemy gets on the bridge, the Artillery will open and rake it with grape;

and as soon as he begins to deploy, the Infantry will throw in their fire and charge with the bayonet. While the Cavalry will also be prepared to charge, and be, with a portion of the Light Troops, ready to follow the enemy up, should he be driven back, if such a measure should be deemed adviseable; at all events, the Troops are thus in a situation either to continue the defence or to commence the retreat again, as may be ordered.

SECTION XXI.

MOUNTAIN WARFARE.

Attack and Defence of Defiles, Passes, &c.

1. When war is carried on in a mountainous country, abounding in strong passes, defiles, &c., or in countries covered with extensive forests, where regular troops are unable to act in masses or in united bodies, but are obliged to fight in detached parties and in loose order, here is the true field for the operations of Light Troops, where they may act with most effect: here their skill, enterprize, and capabilities may be displayed to the greatest advantage; here the activity and individual intelligence of every man becomes essentially requisite; here, in fine, Light Troops are in their proper element; where, under experienced and enterprizing commanders, they may achieve important services, and obtain decisive advantages, even against far superior numbers, if these are less skilful or less ably commanded.

2. The daring conduct and unshaken firmness of the regular soldier, however admirable and of the utmost importance in the open field and in general actions, are but comparatively of little avail when engaged in desultory warfare in countries of such description. Numbers, discipline, and systematic unity of action are rendered almost nugatory when opposed to adversaries endowed with individual tact and desperate intrepidity, united with the nature of the country favourable for such species of loose warfare.

Hence the necessity of Light Troops acquiring the habit of a quick sagacity and perception of employing difficult ground to the highest advantage, when acting in strong mountain defiles, or in very wooded countries, which afford almost impregnable strongholds to belligerent inhabitants, who demonstrate aptitude and skill in Guerrilla warfare.

3. The general principles and rules, stated in the Section on Advanced Guards, &c, are all more or less applicable to the Attack, Defence, or Passage of all Defiles, however difficult, with the addition, that in such countries where the inhabitants are hostile, a strong Rear Guard will likewise be established.

4. In making dispositions to Attack a Defile, where the gorge, with the heights on either hand of it, are occupied by the Enemy, and particularly if he is entrenched, the general rule formerly given must invariably be adhered to; viz., that a demonstration and an endeavour ought always first be made to gain and turn the flanks of the Enemy, by columns or detached parties, either to make real or false attacks, according as circumstances may require, previous to a direct attack in front being attempted. Any other necessary alteration in the disposition, whether one flank only is to be turned, or whether it may be requisite that one or more detached parties should attack a particular flank, will depend on the nature of the ground, and the particular circumstances of each case, and will be determined upon according to the views and judgment of the officers in command. The general principles (founded on experience) remain the same, and he who acts contrary to them, and experiences no misfortune, can only boast of good luck.

ATTACK OF DEFILES.

5. When the gorge has been carried by the Advanced Guard, and the passage to be accomplished, flanking parties will crown and proceed along the heights on either hand, ready to drive off or to out-flank any parties of the Enemy, should they appear; while such elevated points as command the direction of the march should be occupied by detachments to protect the column during its progress, until relieved by the approach of the Rear Guard, when they will follow the column.

PASSAGE OF DEFILES.

6. Should the Enemy take up a fresh position in the Defile, the same measures as above will be resorted to, and the attack renewed.

7. When Skirmishers advance to attack a Defile, which from its particular position cannot be turned, they must endeavour as far as possible to get on the flank of the Enemy, and, by a concentrated fire, drive him from his position, aided if possible by the fire of Artillery, to demolish barricades, &c., while the Supports or a portion of the Reserve may be brought up in column under cover of the fire, to dash in and storm his post.

8. If this succeeds, the Supports may follow up and pursue the Enemy, either extended or not, as the ground permits or circumstances require; and the old Skirmishers may form the Supports.

The Reserve during these operations ought to be kept well in hand, ready to give such prompt assistance as may be required; and also to repel any attack, should the Enemy have succeeded by any detached party to gain its Flank; and such cases have occured.

9. In carrying on a Defensive warfare in such intricate countries, particularly in defending passes, ravines, woods, &c., experienced Light Troops may make the most determined resistance against an overwhelming force.

DEFENDING PASSES.

10. In such operations Riflemen may be employed with great effect, especially in the defence of narrow gorges or sharp angles of mountain passes. Twenty Sharp-shooters, posted on an elevated point of difficult access, may do more injury than several pieces of cannon, (which in such places cannot be moved without much trouble, and in case of a reverse must be abandoned).

Whereas Riflemen so posted would annoy detachments and convoys at a distance where Infantry muskets could not reach them; and on many occasions, at particular places suitable for such purposes, might carry confusion into columns of march, dismount Artillery, and throw Cavalry into disorder, so as even to arrest the march of an Army. Thus in Tyrol the peasants made the French Armies feel the effects of their Rifles. With this murderous weapon, the people, perched on high points, or lying in ambush behind rocky craigs, or in ravines, woods, &c., mowed down their invaders, particularly in the gorges of Brenner and those of Glurnish.

11. When Light Troops in Skirmishing order are retreating, and enter a defile, the Supports will extend and occupy the entrance or gorge. The chain of Skirmishers will, on coming near, rapidly retire under cover of the new line, and form on the nearest best position, either to act as Supports or to extend as Skirmishers again as may be required; or if the Defile continues of some length, or if the Enemy press vigorously on, there may be two Lines or Divisions of Skirmishers and Supports, one to resist the Enemy, and the other posted at the next best position in the rear. By either of these means, the Lines can continue retiring, alternately

relieving and covering one another, and a suitable position for defence can always be calculated upon with a certainty of its being properly occupied. On all these occasions Flank parties from the Reserve will endeavour, if practicable, to act on the Enemy's flanks as occassion may offer. On leaving a Defile and coming to open ground, the movements will be executed as stated under the head of Rear Guards, Section XIX.

Many and various examples might be given of the different methods by which celebrated Defiles, under difficult circumstances, have been forced during the grand European campaigns, but none seems more appropriate at the present moment than that which the recent events in Affghanistan has afforded. However great the gloom which the untoward disaster in that country had occasioned and clouded the glory of our arms in India, the successful forcing of the Khyber Pass has in a great measure tended to retrieve the misfortune and to vindicate the honour of our Indian Army. In order to enable my general readers to judge of the great difficulties, both physical and moral, which the Army had to contend with, the following condensed account of the people, country, and defiles, with the successful mode of attacking the latter, and the overcoming of these difficulties, may not perhaps be here out of place :—

Affghanistan is a mountainous territory, lying between Persia and the Indus, which river forms its eastern boundary. The whole country, being occupied by rugged mountains, only passable by particular and formidable Defiles, presents almost insurmountable obstacles to the movements of a regular Army. Indeed there are, strictly speaking, only two practicable routes to Cabool, the capital, from our Indian empire: the first is from Bombay, by the lower Indus, crossing that river at Sukur by Shikarpoor, through Scinde to Dadur, up the celebrated Bolan Pass to Kuetah, through the passes of Hykulzye, &c., to Kandahar; then by Kelat Ghilzee to Ghiznee, and from thence to Cabool.

The second route is from Delhi and nothern India, by Ferozpoor across the Suttlege river into the Punjaub to Lahore, by Ramnegurh across the Jelum or ancient Hydaspes to Attock, then across the Indus and enter Affghanistan; then by Peshawer Jamrood through the Kyber Pass to Jelalabad, from thence through the Koord Cabool Pass, of melancholy recollections, to Cabool. The extent and magnitude of the difficulties of penetrating into this country is proved by the serious loss of material during our peaceable advance into it in 1839, and in the disastrous attempt to leave it in 1842.

The population is composed of different tribes of various origin, and all of a hard and ferocious nature. The principal of which are:—

The Ghilzies, the ancient sovereigns of the country, who now form the mountain tribes on the west side of Cabool; but the kingdom has passed from their hands:

The Affghans were a powerful and rival tribe; both these conquered Persia, and the last of the Seffis resigned to Mohamed Ghilzee, at Isphan, in 1722. But the Great Nadir Shah rose in 1727 to free his country. He not only drove the Affghans out of Persia, but defeated them in their own country. After this, Ahmeed Khan, a brother of the chief of the Abdaullees, with a large body of the clan, entered Nadir's service in 1737; but when Nadir was murdered in his camp, at Coucha, in 1747, Ahmeed, with all Affghans, abandoned the Persian service. He repaired to Kandahar, collected his victorious troops, and proclaimed himself King of the Affghans. He took the title of Dooree Dooranee, changed the name of his tribe to Doorannee, and died 1777. Thus the Doorannees became the hereditary chiefs, and founded the kingdom: the clan Populzye gave the kings to the Affghans. In 1799, Shah Zeman, the then ruler, ordered Serrafraz Khan, of the Baurikzye tribe, to be slain. This person left twenty-two sons: Fati Khan, one of these, organized revolutions, dethroned Shah Zeman, expulsed Shah Shooja, who, being defeated, fled to India under British protection, and lost the throne of Cabool in 1810. Fati Khan raised his brother, Shah Mahmud, to royalty, with himself as vizier. In fine, these brothers divided the Doorannee empire. Dost Mohamed, one of the most influential, become Khan of Cabool, while another brother became chief of Kandahar.

The Kuzzilbashes are a powerful tribe at Cabool, of Persian origin, and still retain their character of military adventurers. The Ghuzees are a fanatical religious set of Musselmen.

The Kyberees derive their name from the renowned pass: they consist of various tribes, viz., Afreedees, and Orookzyees, &c. Their hill tribes are undiscipled hordes, scarce subject to any rule, and have held the passes from time immemorial. The Khyberees have always made money of their passes, and always take the side of the party that pays them best. Even the renowned Nadir Shah, although a conqueror, when about, in 1739, to enter India with a powerful army, paid £100,000 to the Khybers for the right of the road; nay, even the Affghan rulers, the kings of Cabool, have been in the habit of paying them a subsidy of £8000 per annum for the free use of the passes: but such is their habits of rapine, that they can never be entirely restrained from plundering passengers. In quiet times, they have different stations where they collect tolls; but in times of trouble they are all on the alert: the noise of a single horse's feet resounds up the side valleys and ravines, from whence they rush to intercept the rider. In time of war, it is impossible to pass. They refused with scorn the offer of 15,000 rupees, made to them by Akbar Khan to resist the passage of our re-inforcements coming up to relieve Jelalabad, as being too small; and they opposed the re-inforcements under Colonel Wyld, because he offered them nothing.

They are armed with long jongalls, or with matchlocks, which carry an extraordinary distance, and make use of a wooden perch as a rest to fire from; they likewise carry long knives, swords, and short spears. They are excellent marksmen, and are reckoned good soldiers for the particular service of mountain warfare, being often engaged for this sort of service by other states. They are, however, more disposed to plunder than war, and will fall upon the baggage of the army they belong to if they find it unguarded.

The native chiefs of all the various tribes in Affghanistan, by confining themselves to desultory operations, have acquired great skill and address, and are very effective in that sort of warfare. However irregular their armed bands may be, they pursue a well-arranged and connected system of warfare, and have attained so much tact in their mode of attack and defence, which in such a difficult country, and in almost inaccessible fastnesses, has often foiled the best exertions of disciplined bravery. They seldom venture to meet a regular force in a fair stand-up fight; but if it is on the march, they harass its flanks and rear by firing from every rock or height it passes near. If in camp, they hover round, threatening night attacks, so that an army must be continually on the alert, whether halted or in movement. They, however, can only act with decided advantage against an European army when it has imprudently been placed in such a difficult position that they can bear upon its communications and supplies, annoy its flanks, and cut up its stragglers, or barricade the passes in front: such unfortunately was the case in the recent disaster. Hitherto in our Indian wars our Troops have been expected and have generally been accustomed to carry everything by a *coup de main* against any superior force; but in mountainous districts, abounding in defiles, &c., such mode of warfare cannot be ventured upon. Should a regular Army attempt any such system of operations, it would be exposed to be checked and to imminent defeat, however excellent the general plan of the operations might be, as proved to be the case in the Nepaulese war of 1814, amongst the ridges of the Himalaya mountains. Sir David Ochterlony saw that in such regions the usual style of Indian fighting would not answer; he therefore turned the campaign into a war of posts, out-manœuvered his antagonists, turned all their positions, advanced leisurely, with due precaution, securing every post, thus forming a chain to secure his rear; and also by this means taking care to leave no portion of his force exposed, or liable to be cut off. He thus gained possession of a country as difficult to enter as can be imagined, defended not merely by wild tribes, but by troops far more expert than the Affghans, and who displayed no small knowledge of modern military science.

But the hostility of the natives and the difficulty of the passes are not the only obstacles which an Army has to contend with; there are others of a very serious nature, viz., the extraordinary extremes of climate. In winter, the severity is so great that all the roads are frequently blocked up from November to April; while during summer, the heats are excessive with, at times, a hot scorching wind, named Simoom, which stifle men and animals, and the large tracts of barren waste which diminishes the means of procuring forage for the beasts of burden, as the following will show:—

In the spring of 1839, when Sir John, now Lord Kean, advanced on Ghuznee, although he had no decided opposition to contend with, but only a few bands of robbers who hung upon his flanks, and who only in some slight degree impeded his operations, yet in making his way through the great southern or celebrated Bolan Pass, his army, which consisted of only 15,000 men, required nearly a month to complete the passage. The camels and bullocks died from the inclemency of the weather, or else, for want of forage, were left on the way to be seized by the robbers. The horses perished in great numbers, so that the tents were obliged to be abandoned, and the soldiers were obliged to be put on half-rations, because in this territory no dependence can be placed on any voluntary supplies; so that every contingency must be provided for, and an Army must carry all its resources, and its own provisions for subsistence by the way, along with it. The immense numbers of camp followers, required by the customs and the nature of the warfare of these Eastern countries, is a serious consideration, and much increases the difficulties in a hostile and barren country.

The gallant Sir Robert Sale, having maintained his perilous position at Jelalabad during the whole winter with extraordinary fortitude, while his troops were suffering great privations, it became of vital importance that immediate succour should be sent to his relief as soon as the season would permit. General Pollock therefore determined to attempt the forcing of the Khyber Pass, although the whole of the force destined to act under his orders had not reached him at Peshawer.

The Khyber Pass is one of the most formidable and impenetrable defiles in Asia, considered as a line of military defence. It extends from the entrance near Jamrood in a north-easterly direction on the road to Jellalabad for the space of twenty-eight miles. It is in general very narrow, and commanded on each side by precipitous mountains, so that it has hitherto been reckoned impassable for an Army when the inhabitants had determined to oppose them.

From the entrance to the Fort Ali Musjid, the dell is deep and uninterrupted, (except near a bridge at one point). This fort stands on an isolated hill in a narrow of the Defile, which it completely commands. Beyond this the ascent is somewhat uniform for about seven miles. It then begins to become more difficult, and before reaching Lundie Khan, the head of the Defile, thirteen miles from Ali Musjid, the road stretches along the face of a frightful precipice for nearly two miles, like the Galleries of the Simplon, being cut out of the solid rock, and only about twelve feet broad. Another account says, "After entering the Pass from Jamrood between two high rugged hills, the road ascends up the bed of a dry mullach, or mountain torrent, and then leads along up the side of a hill by a narrow path ; it then continues for three miles up and down over steep rocky mountain-ridges and through narrow ravines, wild and difficult the whole way to Fort Ali Musjid ; after which the Pass keeps ascending for seven miles. At the summit there was a stockade as a post for Troops. The road then rapidly descends, and becomes very difficult, being cut out of the side of the rocks for two miles, and only twelve feet wide. The pass afterwards opens out to about a quarter of a mile broad, and the road becomes tolerably good for eight miles, after which it again closes, and, at the gorge, is not above 100 feet wide, having hills 2000 feet high on either hand. Another difficulty in the passage of this Defile is that the road in many places leads along the confined bed of numerous torrents, which render it particularly dangerous in case of sudden falls of rain.

Forcing Khyber Pass.

General Pollock having made a reconnoissance, found that the Enemy occupied the mouth of the Pass in great force, and had fortified the gorge with a strong breast-work of stones and bushes; that the hills on the right and left were rocky and precipitous, presenting great natural obstacles to the ascent of Troops, and defended by numerous bodies of the Enemy. Hence it appeared, that the most practicable mode of attacking him with advantage was to turn his flanks. To accomplish this the following plan was adopted, and the necessary dispositions made:—On the 5th April, 1842, General Pollock moved from the encampment of Jamrood before day-break in the following order;—

The main Column destined to assault the gorge continued to advance by road, while Columns of Light Troops were detached to the right and left, to gain the heights on each flank of the Pass.

The stores, treasure, ammunition and baggage, were left under a sufficient escort on the road leading from Jamrood. The rear was covered by a Rear Guard under General M'Caskell, consisting of five pieces of Artillery, some squadrons of Cavalry, and detachments of the 9th Regiment, and of the 6th and 60th Native Infantry.

Previous to commencing the attack, the flank Columns were obliged to make a considerable detour, respectively to right and left, in order to enable them to gain the flanks of the Enemy and commence their ascent.

The Right Column, consisting of four companies of the 9th Regiment, four companies of the 26th Native Infantry, and four companies of the 64th Native Infantry, under Lieutenant-Colonel Taylor, 9th Regiment, and 100 Jezailhees, was formed into three divisions of four companies each, with a Reserve under Captain Gahan, covered on their right or outward Flank by a squadron of the 3rd Light Dragoons.

These divisions took up their ground, so as to form different points of attack, viz., one under Lieutenant-Colonel Taylor, composed of two companies of the 9th Regiment and four of the 26th Native Infantry; another under Major Anderson, further to the right and rear, composed of two companies of the 9th Regiment and four of the 64th Native Infantry; the third 100 Jezailhees. Each of these having thrown out a line of Skirmishers with their Supports, so as to embrace the base of the heights and out-flank the Enemy, they all proceeded in this order to the attack. On reaching the foot of the hills, the Enemy's irregulars, who were posted there, immediately opened their fire, and commenced retiring upwards, keeping up a lively fire on their pursuers. Our Skirmishers and their Reserves, notwithstanding the almost inaccessible nature of the ascent, succeeded in driving a large body of the Enemy before them up the hills, which were scaled and crowned in spite of a determined opposition. This being effected, Major Anderson remained to crown and maintain the heights, while Lieutenant-Colonel Taylor descended to his left, to clear the sungahs or entrenchments commanding the entrance to the the Pass, which the Enemy abandoned, and who suffered severely in their retreat. He then proceeded to drive off the Enemy from the positions along the right of the road leading to Ali Musjid, which was finally accomplished, though obstinate resistance was offered at several points, especially at a bridge where the enemy had concentrated in force. Here the out-flanking system was again successfully resorted to. Lieutenant-Colonel Taylor detached two Light Companies to the right, to take the Enemy's position in reverse whilst he attacked them in front; this had the desired effect of forcing their immediate retreat and clearing the bridge. No further opposition was offered by the Enemy on that side, who retreated on Ali Musjid, followed by Lieutenant-Colonel Taylor, who took up a position at a tower on a hill about a mile from that place.

The Column destined to attack the heights on the left of the pass, under the command of Lieutenant-Colonel Moseley and Major Huish, was composed of four Companies of the 9th Regiment, four of the 26th Native Infantry, four of the 64th Native Infantry, and 400 Jezailhees under Captain Ferris. These being duly disposed in Skirmishing order, with a Reserve, advanced to attack the outward heights on the left, which was speedily carried in the most gallant manner, and the summit attained. As this commanded the lower detached hill, at the entrance of the pass, it was soon cleared of the Enemy, by a sharp fire opened upon them. This being effected, Lieutenant-Colonel Moseley remained with the Rear Column to crown and maintain the heights, while Major Huish continued to advance and clear the heights on the left of the road leading to Ali Musjid. He experienced considerable opposition at several points, especially at one of the hills, and at the heights commanding a bridge, where the Enemy made a stand in great force. They were, however, driven from all their posts in the most spirited style, with considerable loss, and followed up by Major Huish, who occupied the last height adjacent to the intended encamping ground of the main Column, near Ali Musjid; during these operations, the Jezailhees were conspicuous in forcing the Enemy to relinquished their strongholds.

While the Flanking Columns were in progress on the heights on either hand, the guns were placed in position, and threw Shrapnell shells among the Enemy, when opportunities offered, and which assisted much in their discomfiture, while they at the same time kept up a well-directed fire on those posted to defend the breast-work at the gorge.

General Pollock, finding the heights all gained and in our possesssion, advanced the main Column and destroyed the barricades at the mouth of the pass which the Enemy had evacuated. It then continued its route under protection of the parties crowning the heights, as all opposition on the part of the Enemy may be said to have ceased on their finding the heights secured. Thus, notwithstanding the determined resistance of the Enemy, aided by the natural difficulties of the route, the Army succeeded in reaching Ali Musjid in the course of the day. This fort was garrisoned by a body of Affghans; they, however, evacuated it when the English Troops came in sight : it was immediately occupied by an irregular corps in the British service. The baggage, &c., covered by the Rear Guard, followed the main Column, and bivoucked for the night two miles within the pass. They both reached Ali Musjid next evening without any loss, not a single baggage animal having fallen into the hands of the Enemy.

After this the Enemy seemed to have lost all confidence in hemselves, and fled. The remaining thirteen miles to Lundie Khan, at the head of the pass, was traversed almost without opposition, and the Rear Guard reached Dakha, eight miles further on, clear out of the pass, on the 10th April.

Thus the Khyber Pass, hitherto believed to be impenetrable in the face of opposition, has been forced; a weak Anglo-Indian army has achieved, in despite of superior numbers, what was never before accomplished by large armies, even when this pass was only defended by an inferior force. This important result has been attained by the steady gallantry, patience, and good conduct of the Troops, skilfully directed by an able General. The effect of this achievement will be felt throughout Central Asia. It must be recollected that Kandahar, Ghiznee, and Cabool have often yielded to the arms of victorious Enemies: the Khyber Pass has never been forced until now.

To prove the necessity of being prepared to resist Flank Attacks, when engaged, even when our Troops are the assailants, the following may be mentioned:—The Khybers, although defeated and driven from the summit of the heights by the right Column under Lieutenant-Colonel Taylor, detached a large force, which stole round a shoulder of the hill, and gained the right flank of the Reserve or covering Column, under Captain Gahan of the 26th Native Infantry.

This Reserve, notwithstanding such an unexpected attack (being prepared for all contingencies), showed a most determined front against an overwhelming Enemy, and steadily moved from pinnacle to pinnacle, fighting every inch of the way, until it safely reached the breast-work on the brow of the hill commanding the descent, and which Major Anderson, with the crowning column, had maintained against numbers. The Enemy, finding all further attempts fruitless, gave up the contest, and disappeared. The Reserve towards evening descended, without molestation, into the pass, and followed the Column.

SECTION XXII.

Defending Woods.

1. FIGHTING in woods forms one of the most prominent and important applications of Skirmishing, particularly in defending them, where Light Troops may be most advantageously employed. The value of such positions are principally as follows:—their great impenetrability, and the advantage of trees to cover the Skirmishers, while all the beneficial peculiarities of other strong positions, such as heights, broken ground, rocks, swamps, &c., may frequently be found more or less combined within them. Further, the circumstance that the defenders are better acquainted with all the localities and internal roads of communication than those who attack. The Enemy must be involved in great uncertainty with regard to our strength, measures of defence, or movements from the impracticability of seeing, all view being obstructed. These, exclusive of other reasons, are sufficient to demonstrate the superior advantage of Skirmishing in such situations; and if it is considered that, when time permits, these local circumstances may be rendered almost impregnable, it may be safely affirmed that, for Light Troops, no more favourable opportunity can possibly offer for an effectual defence than a wood; because, if Troops occupy a position, or fortified camp, the outline of the wood can be rectified at pleasure by cutting off the salient angles, and barricading the re-entering ones, so as to give the whole outline in general a proper form, with entrenched abattis. In chance actions these dispositions cannot be made from want of time; but the roads or entrances may be cut or closed with trees, and every advantage taken of the natural outline and internal circumstances.

The successful defence of a wood principally depends on maintaining the border. To decoy the Enemy into the interior, and then to come to close action, is not advisable unless there offers some entrenchment or very favourable ground for this purpose. Therefore the edge of the wood will be strongly occupied by Skirmishers, and, if possible, supported by Artillery, which, in conjunction with the Skirmishers, will be found to prove eminently effective. The guns should be so placed that they cannot be taken by any sudden onset; and also that they may be enabled to make a secure retreat.

2. In considering the position of the Skirmishers behind trees and fences, if the wood is enclosed, the following advantages appear:—a good shot under such cover, it may be calculated, will have disposed of one or two enemies (if the ground in front which they pass over is at all open) before a third can reach the edge of the wood, so that he may only have one to contend with, and who will be most likely much exhausted by a rapid advance. The Enemy's Artillery will be found to produce more noise than any great detrimental effect; and from his Cavalry there is nothing to fear.

3. The Supports must be placed in such positions that they can quickly render every assistance to the Skirmishers, and be ready to aid in driving back any successful assault of the Enemy. In such cases, when the position is very extensive, and communication along the rear difficult, it is advisable to have local reserves of companies towards each Flank at least.

4. The proper Reserve ought in general only to be employed at the moment of a critical emergency, as it must be ready to move to meet the real attack. The proper use of this body proves the clear-sightedness of the Commander, that he is not to be deceived by the Enemy, but can distinguish false attacks from that which is intended to be the real one. A few intelligent men may be ordered to climb up the tallest trees at different points, to observe and give notice of the movements and principal attack of the Enemy.

5. The main body of the Troops will, if the wood is small, be posted in rear of it; but if the wood is of great extent, it will take a position in the main road. This force is destined to prevent being surrounded, to prevent Flank attacks, and detach parties to look out and resist them; but chiefly to cover those engaged if driven from the wood, or to restore the combat.

6. The salient angles or projecting tongues of a wood are the most dangerous points for the defenders; although at times they may be very advantageous for giving a Flank fire,

or in making Flank attacks on the assailants, still they are to be considered, as they are in entrenchments, the weakest points. Therefore they ought to be well guarded by a sufficient force, and abattis, if possible, which will also afford the following advantages, viz., that if the Enemy attacks a salient angle, he will find a body fully prepared to resist him; and if he commits the mistake to attack a re-entering angle, these Troops will be enabled to take him in Flank.

7. The Skirmishers defending the borders of the wood should be frequently relieved, in order that a continual lively fire may be kept up, which will keep the Enemy in great uncertainty of our intentions, and lead him to believe that our force is superior to what it actually is. As soon as the real attack of the Enemy becomes apparent, the Reserve will move to support that particular point of the chain, and use every possible means to repel the Enemy.

8. Should the Enemy get possession of the edge of the wood, he would acquire an evident advantage; his further movements would remain concealed from us, and his Skirmishers would be nearly on equal terms with our own. The only drawbacks against him consist in his uncertainty of our strength, and of any strong positions we may occupy. This acquired advantage of the Enemy may be diminished should the Reserve, assisted by Flank attacks on him, be enabled to drive him back. But if the maintaining of the wood is still warmly contested, this is the moment the main body should be brought into operation, and on this depends the issue of the action.

9. Should it appear evident that the wood can no longer be maintained, a strong body of Skirmishers from the main body should immediately be sent to occupy the rear edge of the wood in the line of retreat, and the Cavalry posted at a proper distance beyond the wood to cover the Retreat. When these precautionary measures have been established, the Troops will be withdrawn, at the same time preserving the greatest order, if possible, covered by the Skirmishers engaged with the Enemy, who will give ground with more or less rapidity as circumstances may require, until approaching the new line posted at the rear extremity of the wood; when they will dash to the rear through them, so as to clear the front with all expedition, and join the Reserve.

10. By having the rear edge of the wood previously occupied with the Skirmishers, the following advantages are obtained,

viz., that in case it should be wished to re-occupy it, a decided advantage is still in our hands; or that if the Troops are likely to come out of the wood in disorder, they may rally, and order again be restored under protection of the fresh line, and the further retreat properly arranged. Hence it ought to be impressed on the men forming the Chain, as a point of honour, that on their bravery depends the safety of the Troops; and that they must make the most determined resistance to hold the Enemy in check. When the Retreat commences, their Supports will extend in the nearest best position or Cover, under protection of the Cavalry; and the line of Skirmishers that were left in the wood will retire with the greatest celerity.

Attacking Woods.

11. The attack of a wood occupied by the Enemy which cannot be out-flanked, or out of which the Enemy is not to be decoyed, will seldom be accomplished without great loss. To make a direct Front Attack would cost dearly. The assailants cannot know the strength of the Enemy, or in what manner he is posted; and from not being acquainted with the interior of the wood, might easily fall into an ambuscade. The dispositions to attack must be based upon the nature of the outline of the wood or of the ground, and how they are occupied.

12. The salient angles, or any jutting-out points of the outline which are little or not at all exposed to a flank fire, are the true points of attack, should such places offer. But under every circumstance, when it is practicable, endeavour should be made to turn the flanks, and the main attack supported by demonstations or false attacks.

13. In most cases, particularly when acting on a large scale, a line of Skirmishers should first be sent on, in conjunction with a few guns, in order to obtain in some measure a knowledge of the position, and how the wood is occupied. When the real point of attack is decided upon, false attacks, supported by bodies of Troops, must be made at the same time, which may likely mislead the Enemy, and induce him to send on his Reserves too soon in support, and probably to the wrong place; from which circumstance advantage might be taken, by endeavouring, with Columns supported by Artillery, to pass through at other places which he may have left exposed.

The guns will endeavour, with, howitzer or grape-shot, to clear the edge of the wood, or at least to keep the Enemy's Skirmishers in check from acting very vigorously. Shells from howitzers or rockets are the best and most effective. By their bursting, they occasion in a wood great alarm, and may cause the Supports often to change their position; and thus may give rise to disorder; or the rockets may set the woods on fire. But the success, even of this sort of attack, is often doubtful.

14. Each particular attack, whether only a demonstration or in earnest, ought to be made by a weak line of Skirmishers, followed at due distance by Troops in Column. These movements have no other object than to draw the Enemy's attention towards them, so as to distract him, and leave him in doubt which is the real point where the true attack is to be made.

15. When Skirmishers advance to attack the borders of a wood, if the ground is open, they will proceed over it with the greatest celerity, and endeavour to get under some cover at a short range to open their fire. And the earliest opportunity will be taken for the whole line to make a simultaneous dash up, and endeavour to drive the Enemy from the skirt of the wood, and to seize it themselves. If this object is attained, a grand advantage is won, and the assailants will stand nearly on equal terms with the Enemy, in so far that he cannot observe their further dispositions.

16. These attacks of Skirmishers ought however not to be too much relied upon, else the Enemy must be guilty of great cowardice. Columns following up such preparatory attacks may only be enabled effectually to penetrate or attain success under protection of Artillery. The wish to gain a wood with a line of Skirmishers, signifies as much as to ask a question and only to obtain a rude answer.

17. If a salient angle has been gained, more Troops ought to be brought up, and an endeavour made to follow this success by a cautious advance, supported by flank parties.

18. Should the wood be gained, and should the Enemy still occupy the rear edge of the wood, with Skirmishers to mask his retreat, a strong force should be pushed on to drive them quite out of the wood, and then a swarm of Skirmishers thrown out, who, by a lively fire, must endeavour to perplex the Enemy and increase the confusion.

19. If Cavalry are at hand at this moment, a great advan-

tage may be attained, and the Enemy completely routed. But before any attempt is made to pursue, after getting clear of the wood, the Troops should be re-formed in proper order.

Acting in Forests.

20. With regard to wooded countries, such as, for instance, America, covered with extensive forests or but partially cleared along the frontiers, and where a well-understood mode of bush-fighting is practiced, where one can seldom have a fair view of the Enemy, the opponents have a great advantage over the best European Troops, from their knowledge of the nature of the country, which tends to facilitate all their movements.

21. Light Troops, and particularly Riflemen, are peculiarly adapted for service in the North American provinces. From the advantage of their arms being of a superior description, and shorter than a musket, they are handier, and give them an advantage in woods: while their dark clothing and black accoutrements enable them, in a partly cleared country or open grounds, to stand behind the remaining tree stumps, so as at some distance to be scarcely distinguished from them. This was proved in the practice during the former war of 1814, of that excellent Light Corps the Glengarry Fencibles and the Canadian Voltigeurs. These latter, being dressed in a darkish-grey uniform, exactly resembled the stumps and trunks of dead trees; whilst a red jacket could scarcely in any cover be concealed from the lynx-eye of the American Rifleman. The men composing these Corps were mostly all born natives of the Provinces, the Glengarries of Scotch extraction, and the Voltigeurs, French Canadians. These men became excellent soldiers—well adapted for the service for which they were engaged. They knew the woods as well as the Americans, having a knowledge not only of the localities of the country but also of the habits and manners of the Enemy. These men could make their way through the forest where a British soldier would be lost; because, let any one not accustomed to the country penetrate only a short distance into the Bush, as it is there termed, if he happens to turn round or to forget himself, he will scarcely ever be able to find his way out again, much less to proceed in any required direction.

These men were likewise capital shots, being accustomed from their youth to the use of the Rifle practice,—accustomed to the use of the axe,—erecting with celerity abattis, stockades, block-houses, &c. And from many of the Canadians being inured to the hardy life of Voyageurs in the far North West, they were skilful in the management of batteaus, boats, canoes, &c., when required in the passage of rivers, creeks, and descending the stormy rapids.

22. Experience has taught the Indians and native Canadians a few simple rules which may in some measure obviate difficulties to one unaccustomed to these forests, and which by a little practice may enable a person to find his way under such circumstances. In the first place, previous to entering the Bush, if in the day-time, a person should observe the direction of the sun, if it is visible at the time before he enters; or if at night, he should make himself acquainted with the North Star, so as in either case to have a certain fixed point, which will enable him to move in the direction he ought to go, according to the relative bearing such village, out-post, &c., should have in regard to the sun, if during the day, according as the Sun may be in the East, South, or West, viz., morning, noon or evening; or if at night, to the North Star. And all persons, particularly officers on service, should provide themselves with a pocket compass, the variation of which is very inconsiderable in Canada, it seldom varying more than one degree either one way or other.

23. When in the woods, he will observe that the north side of a tree is indicated by the bark being always rougher, and being frequently covered with moss, which the south side is not; that from the great prevalence of westerly winds, the trees have an inclination or list to the eastward, and that the branches grow and spread out more to the south-east side of the tree than to any other.

24. When one has ascertained the direction he ought to take, and in order that he may not swerve from it, he ought to take up the line of three trees or other mark, and continue to prolong it as he progressively moves on, so as always to keep three trees in line in the direction he ought to go.

25. During a campaign in America, the winter alters the style of warfare, because the lakes and rivers, which in summer can only be passed in boats, are in winter frozen over, and passable for Cavalry, Infantry, and Artillery.

26. In any operations during this season of the year, the business must however be done quickly, on account of the extreme severity of the cold.

27. It would be well if the men were drilled to the use of the axe. It is a most useful art in America, where all the natives handle it with great dexterity, and can hut themselves, form abattis, barricades, roads, bridges, &c., with great celerity.

28. Rockets and Shrapnel case-shot may be employed with great advantage to clear the skirts of woods of Riflemen, and also to create confusion amongst his Reserves in the interior or the Bush.

29. It requires much practice in a wooded country to teach European Troops the proper mode of Bush-fighting. Men unaccustomed to forests (particularly if regular Troops of the Line) are apt at first to get bewildered, and consequently not unfrequently expose themselves very unnecessarily to the fire of a concealed and more practised Enemy.

30. They ought cautiously, and with great acuteness, to examine well the ground and cover in front of them, in order to discover the Enemy. If the wood be pretty open, they will probably observe him dodging behind a tree, or lying down behind a fallen one or a stump. If there is brushwood, or a cedar-swamp, he will most likely be kneeling, or standing stooping, to peep through.

31. When engaged, they ought to attend to the Rules in Section IX., viz., always to keep a tree on the left side, and fire to the right of it, so as to be as little exposed as possible, while the comrade may lie down, No. 15, p. 47.

32. Not to fire so much at an Enemy directly in front, but rather at one in an oblique direction, so as to take him in flank, No. 2, p. 43.

33. And when advancing or retreating, not to move directly forward or backward, but to swerve to right or left, as the case may require, while gaining the next cover, so that the adversary, not being able to take a direct aim, must fire with uncertainty.

34. It is very essential that every officer, in command of a detached party, should have one or two experienced guides in such countries, who are acquainted with all the localities, and also with the habits of the Enemy and his manner of fighting: these should be either long-established settlers or trustworthy Aborigines.

SECTION XXIII.

Defending Villages.

1. ALTHOUGH many of the great battles during the grand campaigns of the last war depended frequently on village combats, yet the attack or defence of such places in a general action, where they form principal points in a position, is very different from those required in affairs of out-post for an isolated village, where only small bodies of Troops are engaged. Villages which can easily be surrounded can only be defended for a short time; and Troops who attempted a continued resistance in such places against a superior force would most probably remain prisoners; therefore the serious defence should only be undertaken when the position of a village is favourable, such as its being situated on strong ground, and where its flanks cannot easily be turned, and where it may have a river or open ground in its front or a height in its immediate rear.

2. An officer entrusted with the defence of a village must first acquire an accurate knowledge of its position, and every part of its interior and general outline, previous to making the necessary dispositions. The defensibility and the measures requisite to be taken will depend on the following points:—

3. On the nature of the surrounding country, whether open, or whether it is woody or broken ground, with ravines, which would enable the Enemy to approach close under cover; whether the environs consist of detached houses, gardens, &c., or whether it is commanded by any heights in front or flanks.

4. The position and extent must be considered, whether it is in whole or in part situated on rising ground, or if a rivulet with bridges passes through it, or if there are ravines in front or rear; whether it runs in regular streets, and

which can only be penetrated at certain points, with the nature of the most salient points—if these can flank the re-entering angles, and how far the circumference is proper for the posting of Skirmishers. When all these points are ascertained, he will be enabled to determine whether the whole can be occupied or only a limited part.

5. In either case he will examine the interior, and determine upon such public buildings, churches, or private houses as are best adapted for defence, or so situated at corners of streets as to enfilade them or one another as to afford mutual support, and whether there are any natural obstructions, such as high rocks, heights, banks, ravines, &c., which may be made posts of defence or places for ambuscade, or whether there is time and opportunity for erecting artificial defences; the best position where cannon may be posted, or where the Cavalry can be placed and employed to make sallies; and how the Troops also may be posted outside, when necessary to keep the Enemy in check from turning the flanks, can be supported; how a safe retreat can be effected, and where the Troops can be posted in rear of the village to secure it.

6. When all these points have been maturely considered, and a plan of defence decided upon, the arrangement ought to commence by erecting barricades at all the outlets and entrances, which is best done by abattis with cross trenches; or by barrels, large chests filled with earth, overturned waggons, or beams, &c., at the same time preparing proper passages for sallies, and building up the back doors and lower windows of the outside houses next the Enemy, closing up the windows of the church or other defensible buildings with turf, mattresses, &c., leaving a slit in the middle to serve as loop-holes. Openings must be made to afford free communication for mutual support through garden walls, &c., or through party-walls from house to house when necessary.

7. The following disposition of the Troops may in most cases be found to answer, making due allowance for particular circumstances:—All the troops not necessary for the absolute defence should be placed in position in rear of the village, and detachments on each flank to check the Enemy should he endeavour to turn them; and a party stationed in the church steeple to observe the Enemy, and make such signals as may have been ordered.

8. All the outward houses in the front and flank, the

garden walls on the circumference of the village, and particularly any strong buildings which flank the entrances, must be occupied by Skirmishers, having a due number of Supports and Local Reserves at suitable distances in their rear. The main Reserve will be posted in the market-place, or other central position that may be found most proper, and which will afford free communication with the Supports and Skirmishers.

9. With regard to Artillery, if there are any commanding points where cannon may be advantageously posted, they ought to be occupied; but in general, the greater part may be kept in reserve, and so placed as to command the leading streets, ready to open should the enemy penetrate.

10. Then if the churchyard, church, or other building, offer effectual means for a particular defence, they must be occupied, as very frequently the maintaining of such points is of great importance. The large granary in Essling, which was on the left of the French, obliged them, although already in possession of the village, to evacuate one half of it. The defence of such villages is almost always determined by the bravery of the Troops. The Commander must narrowly watch to discriminate the real from false attacks, and be ready at the right moment to bring up at the proper time and place the Reserve, or a sufficient force from the main body.

11. Care must be taken that the Skirmishers know exactly the positions of their different Supports, with that of the Reserve, and also the main body, which will prevent any confusion; likewise that there is free communication between the different parties engaged, so as to ensure mutual support. The various Commanders, when not pressed themselves, ought, when circumstances permit, to support one another; and should the Enemy penetrate to attack him from all sides, and expel him again. Great energy will be given to such a defence, if the Troops engaged can at times be relieved in succession from the main body. It was owing to this method being adopted that Aspern was so firmly maintained by the Austrians in 1809.

12. Should it appear evident that it will be impossible to maintain the village much longer, or that orders are received to evacuate it, parties should be posted at temporary barricades, erected to enfilade the streets leading to the main line of retreat; and after the Reserve has moved to the rear, the Skirmishers and Supports will be rapidly withdrawn, followed by the covering parties, and all retire out of the village under

protection of a new line of Skirmishers, which will previously have been posted in the best position, close to the village to cover the retreat.

13. If a village, suppose during the retreat of an Army, is to be maintained, and which has been prepared for defence as above detailed, having all the entrances barricaded, except temporary passages for the last Troops posted outside to retire through, and that the force destined for this purpose consists of one Light corps, three Battalions of Infantry, four Squadrons of Cavalry, and six pieces of Artillery, a sufficient proportion of Cavalry and Infantry will be posted in an eligible position in front of the village, in order to keep the Enemy in check, and make him display a large force, so as to give the Troops more time to prepare for defence. And this bold appearance may lead him to suppose that our force is much greater than it really is. Such number of companies of the Light corps as may be requisite will occupy all the most defensible points round the front and flanks, each one having Supports and Local Reserves. The remaining companies and fifty Cavalry will form the Reserve, and be placed in the most favourable central situation; and a party of fifty Cavalry held ready near a sally port, which should be left at whichever point appears most advantageous near the front for this purpose. And if good positions offer at any particular points for Guns, they may be employed. The main body, with all the remaining Guns and Squadrons, will occupy the best position in rear of the village, with a chain of Skirmishers to cover their front;—and if it is possible, two of the Guns should be posted under protection of detachments of Cavalry to command and enfilade either one or both flanks of the village, as the case may admit, and prevent the Enemy turning the Flanks.

14. When these dispositions are completed, the Troops remaining outside will be withdrawn. The Artillery, followed by the Cavalry, will retire first; and then the Infantry, part of which will occupy such buildings as are fixed upon as points of defence. All the others will join the main body in the rear.

15. If possession of the village cannot be maintained, timely preparation ought to be made for the Troops being withdrawn, while those engaged will continue to make the most determined resistance until ordered to retire; and the main body will prepare to receive the Enemy. For this purpose the Infantry will be deployed, and the Guns so placed as to

command the "debouche," while the Cavalry will be posted on either flank. As soon as all the Troops have come out of the village, and have passed to the rear under protection of the Reserves, and if it appears impossible to make further effectual opposition, the Retreat will be commenced.

16. Should it be found necessary to set fire to the village, the easiest means (after all the inhabitants have been made to quit) is to fill the rooms of several houses in different quarters with dry straw, and, at a given signal, set them, with all the barns, stables, &c., on fire. If any of the houses are thatched, a brand applied or a shot fired into it will be sufficient.

17. The application of the above Rules and dispositions for the Defence of Villages, or the following ones for Attack, are equally more or less applicable to castles, entrenched buildings, churchyards, out-works, &c., where determined Light Troops can make a vigorous resistance, and, from being under cover of walls, houses, &c., can fire with great effect.

18. When a village is to be attacked, a Commander ought, ATTACK OF VILLAGES. before laying down a plan or making the dispositions, to consider how in all probability, according to the nature of the place and its localities, the Enemy can or has disposed his forces; and always ought to suppose that he has made the most skilful arrangements; because in such case any mistake he may have made may prove of splendid advantage. If a village is at all accessible, the assailants have the choice of the points of attack. Hence it will facilitate success if the Enemy can be deceived with regard to the true point of attack. To attain this, simultaneous demonstration, or false attacks, may be made in several quarters, so as to keep the Enemy in doubt which may be the real point of attack. But where it is practicable, the chief dependence ought to be on an attempt being made to turn the flanks and surround the village, so as never in the first instance to attack a village in front with Troops in close order until flank attacks with Skirmishers have been made. This, however, cannot always be executed, as when the village has been entrenched in a position, or when such places are flanked or cut off by rivers, lakes, or other great obstacles. Attacks in Villages are in most cases very difficult, owing to the Troops having to act on the large arc of a circle; and the Commander can seldom superintend the whole of the Troops, while the environs are generally much intersected with enclosures, &c., which renders it diffi-

K

cult to conduct them properly. And yet on such occasions it is particularly necessary that the unity of the chain in each separate attack should be maintained from the points of direction of each, which alone can facilitate all commands being properly obeyed. Hence such attacks require experienced Light Troops, particularly if the Enemy has taken the necessary measures for putting the village in a proper state of defence. As there may be many serious obstacles to overcome, even after the outskirts may have been gained, such attacks ought always to be supported by a few guns, not only to cover the real attack, but they may be usefully employed to enfilade the enemy's line of defence.

19. On ordinary occasions the following disposition may serve for an example:—Suppose a detachment, consisting of five Companies of a Light Corps supported by two or three guns, are destined to attack a village, three will attack in Skirmishing order, and two will form the Reserve. Thus one will make a demonstration in front, while the two others will proceed to form the flank attacks, and each of these latter ones will not only leave a portion in support, but also a part as a local Reserve. These will be ready to push on and improve any advantage that may be gained, or to cover the Skirmishers if foiled in their attack. The Reserve will be held in hand in such position where it can be readily brought up to decide the true point of attack, which, during the course of the action, an experienced eye will soon discern, and will endeavour boldly to attain. In all these separate attacks, should a detached house, a windmill, or other favourable cover within a short range of the village present themselves, advantage will be taken of them as posts.

20. Suppose an advanced Guard, consisting of one Light corps, four battalions of Infantry, and four squadrons of Cavalry, and six guns, is ordered to carry a Village; some companies of the Light corps, according to the extent of the village, will be formed in Skirmishing order on three sides of it, and the remaining ones held in reserve, by which means the Commander will be enabled to examine the position of the village and its defences, of which he will obtain a knowledge as the Skirmishers engage, and thus be able to recognize the proper point of attack. He will then attempt to move on the line of the Enemy's retreat, by sending round a detachment of Infantry and Cavalry, so as to inspire him with a dread of being cut off; and will, at the same time by an attack in close

order, endeavour to seize gardens or other cover on the most assailable flank of the village where he judges the real attack ought to be made. Here he will post his Grand Reserves, and commence the general attack under cover of the Artillery.

21. Should the Enemy evacuate the Village, the Cavalry will be pushed on to take advantage of the disorder generally incident on such occasions.

Light Troops Employed in Sieges.

22. Light Troops may be advantageously employed either in the besieging or in the defence of fortresses, or other entrenched places. In defending such places, the Light Troops will generally occupy the advanced posts, and make all small sorties; and will be employed on all occasions to keep the Enemy's tirailleurs in check. In cases where correct shooting is required, such as defending breaches, houses, barricades, &c., and in confined situations not admitting of large numbers being employed, the best shots may be selected to fire and the others to load. They should be posted in all the batteries and lines along the ramparts to protect the Artillerymen working the guns, particularly when loading them.

23. During a siege, they will be thrown out as covering parties in front of the men working in the trenches, to protect them and keep in check any sallies of the Enemy; and in such situations they will lie down when sufficient cover is not to be had. There is no case where correctness of fire is more necessary than when Light Troops are employed during a siege to fire upon Artillerymen on the ramparts of the fortress; because, in order that their fire may be effectual, it is absolutely necessary not only that the balls should enter at the embrasures, but that the soldiers should only fire at the instant the Artillerymen are loading the cannon. This result is only to be obtained by employing skilful marksmen who have confidence in themselves. As this cannot be properly accomplished with the usual Infantry arms, it is evident that Riflemen are fittest for this service; and as in such service they will be within range of grape-shot, and exposed not only to the fire of cannon, but also to that of the Enemy's Sharpshooters protecting their own Artillery, they must endeavour to cover themselves in the best manner possible; and to this end, when practicable, they ought to dig pits (throwing the

earth towards the Enemy) in which they can place themselves and fire correctly at the embrasures by resting their Rifles on the edge. When a fortress is to be stormed or escaladed, the Skirmishers will dash up to the crest of the glacis, and lie down until the Enemy opens his fire on the advancing columns, when they will commence firing, by aiming at the embrasures to disable the Artillerymen, and also at the top of the walls to harass the Troops lining them, so as, by a well-directed fire, to clear the way if possible for the storming parties.

24. On all these occasions the formations and posting of the Light Troops must be regulated by the ground and other circumstances. An experienced Officer will, according to the place, and exigency of the case, devise the most proper measures to be taken.

SECTION XXIV.

Hints on Disembarkation of an Advanced Corps.

1. When an Invasion is to be made in an Enemy's country those parts of the coast will generally be selected which affords various places favourable for such purpose, so that the Enemy may be held in uncertainty which point may be determined upon for the disembarkation of the Invading Army, and thus prevent their assembling a sufficient concentrated force at every point; and the Commander of the Invading Army, in conjunction with the Naval Commander, will usually select that spot which seems most fit for the landing of the Troops, and will admit of Steam Boats or other Vessels of War to approach near enough to cover the landing, and, if possible, clear the shore of the Enemy.

2. The disembarkation of Troops on active service is, on all occasions, a difficult operation, owing to the great subdivision of the men in the number of boats (and the transports being anchored in irregular order), the difficulty of keeping them arranged in any degree of Regimental order in passing through breakers and the usual surf common to all shores, &c., which, (notwithstanding that every possible precaution may have been taken, and however well disciplined or experienced the Troops may be,) occasions a certain degree of disorder, and some confusion is liable to ensue ere they can be disembarked and formed in proper order on the beach. When this delicate operation has to be executed on a large scale, in presence of an active and enterprising Enemy, these difficulties are materially increased, particularly if he offers any serious resistance.

3. The immediate, and perhaps ulterior success after landing greatly depends on the rapidity and regularity with which the men make good their landing and formation.

4. Each company should descend from the ship into the flat-bottomed boat, launches, &c., by files from the centre. Then the right sub-division ranging themselves on the starboard side, and those of the left sub-division on the larboard. And as soon as they reach the shore, those on the starboard side jump out on the right, and those on the larboard to the left, and immediately front, forming company on centre file.

5. The Light Troops to lead, and be prepared to cover the landing.

6. The boats containing each regiment to rendezvous at the head quarter ship, or other point indicated, and to be arranged by companies if practicable.

7. The regiments of each brigade will be dispersed in such order as the Commander may deem necessary according to the nature of the shore, the resistance to be expected, &c.

8. Previous to the Troops quitting the ships, it should be clearly explained to them how they are to arrange themselves in the boats, and how they are to form on touching the shore.

9. The profoundest silence must be observed during the disembarkation, particularly in the act of getting out of the boats. At this trying moment the energies, attention, and zeal of every officer are required.

10. The men ought to land with their arms unloaded; a single musket fired might lead to utter confusion; nor will any one attempt to load until an order is given to that effect.

11. Should the Enemy have moved down to the water's edge in close bodies, the fire of the ships of war covering the disembarkation would, no doubt, either drive them off, or so disorganize them as to enfeeble their power of attack, and render any position their Artillery may have taken up untenable.

12. Should he only have posted a chain of Skirmishers on the nearest broken ground commanding the beach, it would be advisable to detach a battalion or small body to right and left, with orders to take post in close column if the ground be open, in order to secure the Troops in the act of disembarking, from an attack of Cavalry; and at the same time a line of Skirmishers with sufficient Supports should dash rapidly forward and endeavour to dislodge the Enemy's Sharpshooters, so as to take their galling fire off those still disembarking, as likewise to dispossess the Enemy of, or at all events drive back any Artillery bearing on the spot of disembarkation. At the same time due caution should be ob-

served that the line of Skirmishers are not drawn in too far, or into more open ground after the retreating tirailleurs, when they would be exposed to a body of Cavalry dashing suddenly in amongst them and driving them back in disorder.*

13. Attacks by Infantry in ambuscade will generally be thwarted if the Skirmishers and those thrown out to protect the flanks keep a good look out and search all suspicious places.

14. Should this attack upon the Enemy's advanced Troops prove successful, no time should be lost in instantly seizing upon the best position the ground affords. Every rock, sandhill, inequality of ground, or tree, should be turned to account; and all houses or buildings in the line of defence, or which commands the disembarkation, should be immediately occupied, the doors barricaded, and loop-holes made in the walls, and defended to the last extremity. Should the Enemy return with reinforcements to endeavour to recover the ground they had been driven from, the importance of retaining possession of, and maintaining such ground and buildings in order to cover and protect the further landing of the Troops and Field Artillery, and their advance to support those in the position gained, is self-evident.

15. The details here given have principally reference to Troops landing to form the advanced corps of an army when no very serious opposition is expected or offered.

16. In the event of the Enemy being posted in such strength and numbers as to render it necessary that a combined attack should be made, the main body would act under the dispositions of the Commander-in-chief.

* See Lt. Col. Leach's work.

APPENDIX.

ABSTRACT OF GENERAL RULES FOR SKIRMISHING.

Extending.

To Extend Halted.
1. To extend on the spot, from a Right—Left—or Centre—file; the named file remains fronted, all the others face from it, and proceed straight forward, each file in succession halting and fronting when it has gained the distance ordered. (See Part V., Section III., Nos. 1 and 2, page 261. H. M. Reg.)

On the March.
2. To extend on the march while advancing or retiring; the named file, whether a flank or a central one, will move on in quick time; all the others will make a half turn outwards, and proceed in double time, extending; each file in succession on gaining the named distance will front, turn, and take up the quick step. In this extension the files must look well up to their leading file. (Page 266, No. 11.)

Skirmishers continue to Advance.
3. When Skirmishers are thrown out to cover the immediate advance of a line or column, the chain, after extending, unless otherwise ordered, will continue to move on until further orders are sounded.

Skirmishers to Halt.
4. If thrown out to cover a deployment, they will, unless otherwise directed, in general after extending, occupy the best cover at due distance in front, to protect the formation and wait for further orders, whether to fire, advance, retire, &c.

On Actual Service.
5. But in the field, when a certain position is to be occupied in the line of defence, they will of course halt in it and wait further orders; or if covering a body about to advance to make an

attack, they will likewise, after extending, wait for orders so as to move in combination with it.

Distance.

6. Six paces is the regular distance, unless the Commander names any particular number, as the object in view may require. (No. 3, page 262.)

When a certain extent is to be occupied.

7. When a certain extent of ground is to be occupied, the Officer will point out objects at each extremity of the space to be covered. The Skirmishers will conform to the object in view, and divide the distance as equally as possible during the march. But if the number of paces taken are insufficient or too many to cover the required extent, the Skirmishers must not correct it by shifting about, but will wait for orders, and it will be done from a named file if halted, or by increasing or diminishing the distance gradually, if on the march. (No. 3, page 262.)

Correcting Distance.

Increasing Distance.

8. When in extended order, if the distance is to be increased, on the sound—" Extend,"—Skirmishers will take one-half more distance than they have already got, from the centre or any named file as may be ordered. (No. 5, page 258.)

Chain to overlap the Flanks of Troops.

9. When covering a body in line, whether it is halted or in movement, the chain of Skirmishers must always be prolonged beyond its flank, so as to overlap and protect them. (No. 13, page 274.)

Flanks to be secured.

10. The flank of a line of Skirmishers should always be secured, if possible, by resting them on some strong points, and a double file may be thrown out from each to keep a look-out.

Securing Cover.

11. After extending, Skirmishers will on all occasions instantly get under cover; each file will seize every advantage of ground, whether halted or in movement, and look out for an object to fire at. In occupying the edges of heights, backs of fences, &c., they will follow their direction however irregular, but must be careful to fire clear of one another.

Rules for Skirmishing. 203

Point of Direction.

12. The centre of a line of Skirmishers (which is usually marked by a Sergeant) is considered the point of direction, unless any other is named. Although too accurately dressed lines are not required, still Skirmishers must, by a glance of the eye, avoid losing distance, or getting too far before or behind the file next them towards the directing point, whether a central or a flank one. Hence they will regulate their movements so as to preserve distance, to whichever hand that point may be. (Pages 262, 264, 269, 271.)

13. Light Troops, whether acting as Skirmishers, Supports, or Reserves, will—"Trail Arms"—when ordered to —"March,"— and —"Order Arms"— when halted, without any word of command. (No. 13, page 266.)

To Close.

14. When the —"Close"— sounds with a distinguishing G, the men, if halted, face to the point directed, and close to it. If on the March, whether advancing or retreating, the named file moves on in Quick Time. All the others make a half turn towards it and close in Double Time. (No. 13, page 266.)

Chain Advances or Retires.

15. If the whole line of Skirmishers are to advance without firing, on the sound to —"Advance,"—the whole step off in Quick Time. If the chain is to retire, on the sound to—"Retreat,"—the whole go to the left about, and retire together, rear rank loading, and in both cases will preserve distance and dressing from the centre, or other point of direction that may have been named. (Nos. 8, 9, page 264.)

"Halt."

16. The —"Halt"— annuls every previous sound except the —"Fire;"— therefore if Skirmishers are performing any manœuvre, or are firing, advancing, or retreating, if the —"Halt,"— be sounded, they will immediately stop, but continue to fire. (No. 4, page 257.)

Firing.

To Fire Halted.

17. When a line of Skirmishers is to fire halted,

on the sound to —" Fire,"— the whole, if not already kneeling, will drop on the right knee: the rear rank men disengage a little to the right. Each file takes advantage of trees, rocks, &c.; and where the cover is particularly good, one or two files may occupy it. The front rank men fire and load, then the rear rank, and so continue, observing the general rule that both men of the same file are never unloaded at the same time. (No. 6, page 263.—No. 9, page 273.)

Not to Fire at Random.

18. Skirmishers are never to fire at random; but each man will select his particular object, and take a steady aim, so as to fire seldom and with effect.

Various Positions.

19. In the field, Skirmishers, whether halted or in movement, may fire and load either standing, kneeling, or lying, as the case may require. Should the ground not afford cover, the men who are unloaded will keep their file leader between them and the Enemy while loading. (No. 3, page 269.)

To Load before Advancing.

20. They will always, when practicable,— " Load,"— under cover; and, if advancing, load before moving on; or, if retiring, after falling back, unless ordered to do so on the march; and even then, when any difficulty is experienced, the men, if advancing, may halt to load, and then run up to their file leaders. (No. 9, page 273.)

Two Men to act in concert.

21. When Skirmishers fire, either in advancing or retreating, the two men of the same file must always act in concert and never separate, so as at all times to be ready mutually to support and protect one another, and only fire alternately. (No. 2, page 269.)

Firing Advancing.

To Fire Advancing.

22. When Skirmishers are to fire while advancing, they will push on in a general line. If halted when the—" Advance and Fire "—sounds, they will, on ordinary occasions, proceed as fol-

lows:—The front rank man of each file fires, steps to the left, and moves on, loading on the march. When finished, he gives the word—"Ready"—to his comrade (who had continued advancing in a threatening attitude) to fire; after which the rear rank man will move on, &c.; they will always, when advancing, pass to the front by the—"Right"—of their file leader. (Nos. 4 and 5, page 270.)

<small>When engaged in Cover.</small>

23. But in the field, when engaged in cover, after both are loaded, one man will run on to gain the next tree or other object of shelter, while his comrade protects his advance by aiming at the Enemy, until the cover is secured, when he will close up to the one in front, when both may continue firing alternately as many shots as circumstances and the general movements of the line may sanction: and then, after loading, proceed to push on again in the same manner; at the same time, by an occasional glance, they will keep in view the file next them towards the point of direction, so as to preserve distance and prevent any break in the line. (No. 5, page 270.)

<small>In Pushing on.</small>

24. When the nature of the country is such that the line of Skirmishers must push on from one cover or position to another, they will, on leaving one, gain the next with the greatest celerity, so as never to stand exposed in any intervening open space, and will only fire on gaining such position, where they fire as many shots as circumstances seem to require; and then, after loading, will proceed to make another onset. Thus, on leaving a hedge, they will dash across a field to the next fence. (No. 4, page 270.)

<small>Simultaneous Attacks.</small>

25. In cases where the Enemy, posted in certain positions, is not shaken by our fire, nor seems disposed to give ground, a simultaneous dash up may be made to drive him off and seize his line of defence. In like manner, when a Line advancing approaches the edge of a wood, the crest of a height, or any cover, although the Enemy

Rules for Skirmishing.

may not appear, but where it may be apprehended he may be concealed, a similar rush may be made.

When Firing Halted if the "Advance."
26. When a chain is halted and firing, if the —"Advance"—be sounded, the man of each file who is loaded will move on in a threatening attitude, followed by his comrade, both proceeding as directed. (No. 18 or 19.)

When Advancing, if the "Fire" be sounded.
27. When a line of Skirmishers is advancing but not firing, if the—"Fire"—be sounded, and if in cover, the rear rank man will kneel down and aim at the Enemy, while the front rank moves on to gain a cover, &c. as directed. (No. 19.)

When Advancing and Firing, if the "Halt" be sounded.
28. When a line is advancing and firing, if the —"Halt"—be sounded, the files will get into their places, kneel down under cover, and continue their fire.

When Firing Advancing, if "Cease Firing" be sounded.
29. When firing and advancing, if —"Cease Firing"— be sounded, not a shot must be heard. The men complete their loading, files get into their places, and the whole continue to advance.

Firing Retreating.

Retiring by alternate Ranks.
30. When a line of Skirmishers is halted and not firing, if the—"Retreat and Fire"—be sounded, and that they are to retire by alternate ranks, the front rank men fire and proceed to the rear, loading on the march. When finished, or when a good cover offers, they will halt, front, and kneel down, ready to protect their comrades. The rear rank men, when they find by a glance to the rear that the ramrods of the other Files are working, will—"Fire,"—and retire in Double Time, passing by the proper left of their comrades, when they will take up the Quick Step, commence loading, secure cover, halt, front, &c. (No. 11, page 265.—No. 6, page 271.)

Retiring in General Line.
31. But in the field, or whenever the country is wooded, broken or enclosed, it will always be advisable to Retire in a general line. On the

Rules for Skirmishing.

sound—"Fire,"—the front rank man will fire, run back to the next best tree or cover in the rear, and load; while his comrade will remain kneeling, and aiming at the Enemy, until the one in the rear is ready, when he will fire and retire, passing beyond his comrade to the nearest cover, &c., and so continue. (No. 11, page 265.—No. 6, page 271.)

Retiring from one Position to another.

32. If the whole line, in retiring, must pass from one cover to another, Skirmishers will only fire from behind the position they have gained; and in such cases, on leaving a cover, should the ground be open, they must pass it with all expedition, before coming to a stand at the next cover. (No. 8, page 273.)

Retiring by alternate Lines.

33. If the Retreat is conducted by—"Alternate Lines,"—the Supports will be extended behind a fence, &c., and the former Skirmishers will retire through them, and either gain the next fence and line it as Skirmishers, or form up in Supports, as may be ordered; and so continue retiring by alternate lines. (No. 6, page 272.)

When retiring, should "Fire" be sounded.

34. When the whole line is retiring together, if the—"Fire,"—be sounded, the whole will face about and kneel down. The front rank men fire and retire, &c., and all proceed as directed from the Halt. (Art. 27,—R. No. 11, page 265.)

Firing and retreating, should "Halt" be sounded.

35. When firing and retreating, if the—"Halt" —be sounded, the men next the Enemy will stand fast, (or face about, if not already fronted in that direction). The other rank will close up, Files will kneel down under cover, and the whole continue firing. (No. 12, page 266.)

"Cease Fire" when retiring.

36. When firing and retreating, if—"Cease Firing"—be sounded, the men next the Enemy will instantly retire in Double Time to the rank in rear, which will rise up and face about. The whole will take up the Quick Step, and continue retiring. Those that are unloaded will load on the march.

Rules for Skirmishing.

"Advance" sounds while Firing and retiring.

37. When firing and retreating, if the—"Advance"—be sounded, the Skirmishers will make a momentary halt, fronting towards the Enemy. The Files in rear close up and the men who are loaded will be the first to advance, and fire as directed. (No. 19.)

38. If the line is retiring without firing, and if the—"Advance and Fire"—be sounded, they will face about, and proceed as directed. (No. 19 & 20.)

Advancing and Firing, if "Retreat" sounds.

39. When advancing and firing if the—"Retreat"—be sounded, the men who may then be in front will be the first to retire (firing, if loaded), pass to the rear, secure cover, &c., as directed for firing and retiring.

40. When retreating, Skirmishers after firing go to the right about, and when at due distance in rear, will front by coming to the left about.

Not to collect in Groups.

41. Skirmishers must on all occasions, and particularly when firing, whether halted or while advancing or retreating, be careful never to collect in groups.

Parade, Field-day Firing.

42. When experienced Skirmishers are in action, each only fires as objects and fair opportunities offer; but it frequently happens on a parade field day, that when a line of Skirmishers are to fire, the whole at once throw away their fire, by firing all together. Hence long intervals usually intervene between each round, and the fire, even for show, is but ill sustained. To prevent this the French, in the time of peace, number the Files by threes from the right, and every third alternate File only fires at the same time. On the sound to—"Fire,"—all the No 1 Files fire, then the No. 2 Files, then the No. 3 Files; then the No. 1 again, and so on, so that there is no intermission; but of course this method ceases when they take the field on active service.

Firing while taking Ground to a Flank.

43. When a line of Skirmishers takes ground to a flank, firing on the march, suppose they are

faced to the left, on the sound to—"Fire,"—each front rank man halts, faces towards the Enemy, and fires. The rear rank man moves on obliquing sufficiently to the right to get on the line of defence; and when at about 12 paces, more or less as cover may offer, he will face towards the Enemy; and when his comrade, who will follow loading on the march, reaches him, he will fire. The front rank man then proceeds on, &c., and so on alternately. The Supports also face and move in the same direction. (No. 16, page 267).

44. If the—"Halt"—be sounded, the whole halt front towards the Enemy, kneel down, and continue firing.

45. If the—"Cease Firing"—be sounded, all instantly cease, Files getting into their places, and move on gaining ground to the flank.

46. If the—"Advance"—or—"Retreat"—be sounded, they will in either case front, and then proceed to advance or retire, as the case may be.

Change of Front.

47. On ordinary occasions, when a line of Skirmishers changes front, or if a wing is thrown forward or backward, they will proceed as directed, as the case may be. (Page 268, Nos. 18, 19, 20.)

Keeping up the Fire.

48. But if a wing is to be thrown forward, while the line is firing, they may proceed as follows:—Suppose the Right Wing, on the sound —"Right Shoulders Forward,"—the left flank file will face, and the two next files will be dressed up into the new direction. All the others will conform, each in succession taking up distance and dressing from the inward or halted flank, and open their fire as they arrive in their places.

Wing thrown Backward.

49. When a wing is thrown backward, the inward file will be faced, and one or two files dressed back into the required direction. All the others will face about and move to the rear; and each file, on gaining the new direction, will halt,

front, kneel down, and open their fire. If retiring by alternate ranks, the rank whose turn it is to retire will, after passing the line in its rear, bring forward their shoulders as may be directed, and then halt, front, and load. The next rank, after firing and passing the halted one, will do the same, and so continue. (No. 19, page 268.)

50. If the—"Halt"—be sounded, whether in throwing a wing forward or backward, the whole halt; files get under cover and continue firing.

51. If—"Forward"—be sounded, the whole advance or retire, as the case may be, direct to the new front.

Skirmishers Recalled.

Skirmishers recalled.
52. When Skirmishers covering a battalion or brigade are recalled, they must clear the front as expeditiously as possible, and will adopt that mode to run in in such direction as will soonest enable it to—"Fire"—or—"Advance."—(Page 280, Sec. IV., No. 27.)

Close.
53. If by the—"Close,"—they will proceed as directed, Reg. Part V., Sec. II. (No. 6, page 258.)

Assembly.
54. If by the—"Assembly,"—and if covering a battalion, they with their Supports will rapidly retire generally, by both flanks, so as to keep wide of the battalion; but when covering a brigade, the central Skirmishers will make the best of their way through the intervals between regiments; and the outward ones only pass to the rear by the flanks of the brigade, when they will form up and proceed to such point or flank as may be directed.—(No. 6, pages 258, 259.—No. 26, page 279.—No. 27, page 280).

55. Should Skirmishers be driven in close to the line, they will in such case, when ordered to withdraw, get through it as they best can, two or three files opening at intervals to let them pass. On some occasions they may be made to lie down; and when the lines or columns pass over them,

they will close, and move to such point as may be directed.

When a Battalion forms Square. 56. When a company is covering a battalion which forms Square, the Skirmishers will take the most direct way to the rear, opening well out so as not to run across its front; and then close up and form the rear face. But if not recalled, they may close to whichever flank offers the best position and form Square. Or if there are walls, ditches, &c., at hand, they may occupy such cover so as to take the Cavalry in flank. (No. 28, page 280.)

Squares of Skirmishers and Supports. 57. On parade field-days, when two or more companies are covering a body manœuvring which form Square, the Skirmishers, if not recalled, will fall back on their several Supports, and form Squares with them.

58. But on service, if not recalled, on the appearance of Cavalry, the —" Alarm "— will be sounded, followed by the —" Assembly,"— or — " Form Square."—The Skirmishers, if there is time, will retire and form Squares with their Supports. But if there is not sufficient time, they will **Rallying Squares.** form Rallying Squares on their respective Officers, each wing or company, as they best can, closing on its own centre. Or if there is cover, such as copses, garden walls, &c., the whole, or such portions as best can, will make for it, and aid the Square of Supports by a cross fire. (No. 31, page 281.—No. 33, page 282.)

59. On all occasions, when Squares are formed by Troops in skirmishing order, whether by the Skirmishers alone on their own ground, or united with their Supports, the Officers will take care to choose good positions, and that they are so placed in echellon as to fire clear of one another, as well as of the battalion Square in the rear. (No. 33, page 282.)

Squares Reduced. 60. These detached Squares are usually reduced by the—" Skirmish "—or by the— " Assembly."—If the chain of Skirmishers is to be again re-established on the line of defence, the —" Skirmish"—will be founded; on which the former Supports, as being a fresh body, will dash out, extending on the march, and cover the ground occupied by the former Skirmishers, while the old Skirmishers will remain and become the Supports, unless in the case when Rallying Squares have been formed by the Skirmishers, when they of course will resume their former positions.

Detached Squares Recalled.

61. If Squares, whether composed of the Skirmishers alone or of these united with their supports, are to be reduced by the—" Assembly" —and recalled, the Squares will break, and run in in dispersed order by the flank of battalions, or through the intervals of regiments in brigade.

Sudden Rush of Cavalry.

62. In case of a sudden rush of cavalry, the men should be taught to defend themselves as the case may require, whether individually or to unite in small bodies: in the first instance, by two or three nearest files to one another getting together back to back, and these parties again joining together so as to form larger ones. And to face the danger with coolness and courage, those that can shift for themselves by getting up banks, behind trees, or can gain the nearest fences, or other cover, will do so, and from behind these keep a well-directed fire. They will at the same time take the utmost care to fire clear of one another. (No. 35, page 282.)

Double Files or Chain.

63. A line of Skirmishers in extended order should be practised to form double files or chain, viz., on the sound—" Form Chain,"—the left files face to the right, and close upon the right files; on the sound—"Extend,"—the left files face to the left, and resume their former places.

64. In some situations this formation may prove useful, particularly when there is any apprehension of Cavalry: if practised occasionally previous to exercise, it teaches young soldiers to recollect whether they are right or left files.

COMPANY DRILL.

The following example of Drill Practice for a Light or Rifle Company includes the most necessary movements and Bugle sounds, and likewise affords to each sub-division an opportunity of practising alternately the duties of Skirmishers and Supports :—

PART I.

Suppose one company in Line, or in column of sub-divisions right in front,

No. 1, On the sound to—" Skirmish,"

The right sub-division, or No. 1, dashes out, extending on the march from a Right, Left, or Centre file, as may be named by the Commander;

2. The left sub-division, or No. 2, remains in support, and will take ground to whichever hand it may be necessary, so as to bring it in rear of the centre of the chain;

The company being thus in Skirmishing order, the following movements may be sounded at due intervals :—

3. Advance and Fire—*Cease Firing**—*Fire**—Halt—Cease Firing;

4. Retreat and Fire—*Cease Firing**—*Fire**—Halt—Cease Firing;

5. Alarm—Form Square—(Skirmishers retire in Support)—Fire—Cease Firing ;

6. " Skirmish"—on this sound the left sub-division, or No. 2, which had been in support, dashes out and extends; the old Skirmishers, or No. 1, remain in support.

All the above sounds and movements will be repeated over again, so that each sub-division will thus practise the duties of Skirmishers and Supports.

---

* The sounds in Italics marked * are inserted, to accustom the men to stop firing when required, or to open it quickly when necessary. They may be omitted at pleasure.

PART II.

The right sub-division being again the Skirmishers, and the left one in Support,

7. Line to left, on the sound—" G.G.G. Form Line,"—the chain of Skirmishers wheel up to the left; the support or left sub-division brings Right Shoulders Forward; Advances extending, to prolong the line of Skirmishers to the left. The whole company is now in one line.)

8. Take ground to a flank, firing on the march—Cease Firing—Halt.

9. Advance—Fire—Cease Firing—Alarm—Form Square —(Rallying Square on centre)—Fire—Cease Firing.

10. Skirmish—the whole company extends again to right and left on their own ground.

11. Fire—Advance—Right Shoulders Forward—Forward —Halt—Cease Firing.

12. Line to Right, on the sound " G. Form Line,"—the right wing of the line of Skirmishers wheels up to the right in extended order; the left wing closes to the Right, forming line to the Right, and becomes the Support.

13. Advance and Fire—Left Shoulders Forward—(Supports conform)—Forward—Cease Firing.

14. Relieve Skirmishers—(*Halted*).

The left sub-division being now the Skirmishers and the right one in support, all the movements of Part II. will be repeated, only recollecting in this example to form line to the right, when the right sub-division is in Support; and to the left when the left sub-division is so. But in Field practice, a new line of Skirmishers may be formed to either flank without any reference whatever to which sub-division is the Support or Skirmishers.

PART III.

The right sub-division being now the Skirmishers, and the left one in Support as before,

15. Retreat and Fire—Relieve Skirmishers—(*Retreating*) —Halt—Cease Firing.

16.—Reinforce Skirmishers to the right; on the sound— " G. Reinforce,"—the Support or right sub-division dashes up, inclining to right, and extends to prolong the line to the right.—(The whole company is now in line.)

17. Retreat—Fire—A wing thrown back, by each rank alternately bringing Shoulders Forward—while retiring—Halt—Cease Firing.

18. Diminish Skirmishers—Left Sub-division—Recalled—(runs to Rear and forms Support.)

19. Advance and Fire—Relieve Skirmishers (*Advancing*) Cease Firing—Halt.

The Left sub-division being now the Skirmishers, the movements of Part III. will be repeated when these are finished.

20. The—" Assembly"—Skirmishers run back and form on the Support, in Line or in column of sub-divisions, as may have been ordered.

21. After the men are well versed in Common Drill and Bugle Sounds—no particular routine of movements should be long continued: diversified movements are requisite to accustom the men to be quick and intelligent, and ready to act according to circumstances on any emergency.

22. Two companies may perform the same movements, one acting as Skirmishers and the other as Support—or, the two companies may be thrown into Skirmishing by each leaving a sub-division in support.

23. If from Line, the two centre sub-divisions may advance, extending to right and left from their inward flanks, while the outward sub-divisions move to each flank to form the Supports.

24. If from Column—the companies may disengage to right and left—by Threes—by Filing—or Diagonal March,—and when clear, the inward sub-divisions—Front—Turn—and Advance—extending from their inward flanks. The outward ones continue moving towards the flanks to form the Support.

25. Four companies may act in like manner, by two extending, and the others forming the Supports; or two may remain in reserve, and two act in Skirmishing order, as above.

26. If there are only three companies, one may remain in Reserve in column of sub-divisions, and two may act in Skirmishing order.

27. When the men are—" au-fait "—in all the various light movements in open ground, they should be practised in woods—broken and enclosed grounds; the Officers will pay attention to prevent looseness or disorder;—that the rules

for Skirmishing are properly adhered to, and that the men show intelligence in taking advantage of every variety of cover.

28. Two or more companies should be made to oppose one another in practising the attack or defence of positions—woods—villages—passage of bridges—defiles, &c., &c., and then out-post duties—pickets—patroles, &c.

29. The men should be practised in learning to judge distances. The men, being in squads, a certain hedge—wall—row of trees—crest of a height—or opposite bank of a ravine, &c., &c., may be pointed out, and each one in rotation asked the distance, and the number of paces he may guess it to be marked opposite to his name. It can then be measured by a pacing-stick, or the men may be marched on the object, the Serjeant counting the paces, so that each man will practically find out how far he was right or wrong.

The French sometimes send out a man in front with a light target, which he plants at uncertain distances, varying from 50 to 300 paces, and the men are called upon to name them.

www.ingramcontent.com/pod-product-compliance
Lightning Source LLC
Chambersburg PA
CBHW031140160426
43193CB00008B/201